D1582668

Growing Soft Fruit

Contents

NOTE: The chapters are in the order in which the fruits ripen.

Conversion Table

Metric and Imperial Equivalents

Imperial		Metric	Metric		Imperial
1 in.	=	2.54 cm	1 cm	=	0.39 in.
1 ft	=	0.3048 m	1 cm	=	0.033 ft
1 yard	=	0.91 m	1 m	=	3.28 ft
1 mile	=	1.61 km	1 km	=	0.62 miles
1 sq. ft	=	0.092 sq. cm	1 sq. m	=	10.764 sq. ft
1 gal	=	4.54 litre	1 litre	=	0.22 gals
1 oz	=	28.34 g	1 g	=	0.035 oz
1 lb	=	0.45 kg	1 kg	=	2.20 lb
1 ton	=	1.01 tonnes	1 tonne	=	0.984 tons
x°F	=	$\frac{5}{9}(x-32)$°C	y°C	=	$(\frac{9}{5}y+32)$°F

Metric abbreviations

cm	centimetre
m	metre
km	kilometre
ml	millilitre
g	gram
kg	kilogram

1 An Introduction to Soft Fruits

Why grow your own?

With the high cost of fruit in the shops and with so few varieties available—market growers cultivate only those which provide them with a high yield to compensate for the ever increasing costs entailed in the picking and marketing—it is necessary for those with space available to grow as many soft fruits as possible so as to give an all-year supply and to obtain a wide selection of the best varieties. By growing one's own, the fruit will be enjoyed in fresh condition, at its best, not having deteriorated as so often happens when it is sent on long journeys to the shops or wholesale markets causing much of the flavour (and juice) to be lost before it arrives. Again, it will be on hand almost throughout the year if the surplus is put in the deep freeze, which is now to be found in many homes (where it is the most valuable possession), always providing a tasty meal with meats and choice vegetables, also placed in the freezer when at their best and cheapest, and 'sweets' of fresh fruit salad or of choice fruits which can be used in tarts and flans. What is more, there will be delicious food available whatever the weather and without travelling a distance to obtain it. By using the deep freeze unit, nothing is wasted; the fruits are frozen when at their best and this is how they leave the freezer. There is no need for additives to help them keep their colour and quality and there is no need to partly cook or blanch them as with vegetables. They may be frozen as they leave the plant and will retain their peak condition for a year or more.

Also, the soft-fruit grower may wish to consider selling his surplus fruit or, indeed, to grow the fruit with this in mind. Providing the fruit is well grown and well presented, the grower should have no difficulty in finding outlets for the produce and can expect a good return on the investment (see Chapter 11).

Soft fruits are those which grow on bushes or plants; fruits which grow on trees are known as top fruits, e.g. apples, pears, plums and cherries. Tree fruits is perhaps a better name for top fruits, whilst soft fruits may be best described as bush fruits.

New introductions

During the past twenty years, the introduction of new varieties has revolutionised the growing of all soft fruits. Until the Second World War, there were few reliable varieties. For example, there were only two well-known strawberries: Royal Sovereign and Huxley's Giant, both of which fruited over a period of about three weeks and if the weather was then cold and wet, the fruit decayed before it could ripen and the plants suffered from mildew and botrytis. Late frosts, too, could wipe out a whole plantation overnight with the commercial grower made destitute unless he (or she) had other crops, such as chrysanthemums and winter vegetables, to follow. Much the same occurred with raspberries. Here again, there were only two reliable varieties: Lloyd George and Norfolk Giant, though the latter variety bore its blossom much later and usually missed the frosts. It could, however, be spoilt by wet weather for raspberries, of all fruits, are most troubled by damp conditions. To guard against this, commercial growers planted Careless gooseberries, the canner's favourite, but in most years returns were small.

How different the situation is today. Soon after the

War came Mr Boyes' strawberry introductions from Cambridge and these were soon to revolutionise the growing of this most popular of all fruits. Cambridge Early Pine, resistant both to frost and mildew and cropping well in light soils, ripens its fruit at least a fortnight before Royal Sovereign and crops at the same time. Cambridge Brilliant makes a compact plant with little foliage and so is ideal for cloche or frame culture. What is more, it bears heavily in soils with a high lime content, being the only strawberry to do so. Many new introductions, some from the Continent, enable the cropping period to extend from early May (under cloches), through the mid-summer months, until well into August when Cambridge Rearguard and Late Pine are at their best. But even then, the strawberry season has not finished for still to come are the autumn-fruiting varieties: Hampshire Maid, La Sans Rivale, Gento and St Claude. The plants should be de-blossomed until early June and will begin to fruit late in August and continue (under cloches) until Christmas. By planting for succession it is therefore possible to obtain strawberries from May until the year end, and by deep freezing any surpluses, frozen fruit will be available until such time as the next season's crop begins to ripen.

But that is not all. Strawberries can be grown where there is no garden at all, perhaps only a few yards of ground such as provided by the sunny courtyard of a town house, or a terrace verandah. Here, they may be grown in tubs and barrels, in plant pots or in the new Tower pots. These are plastic pots which are about 6 in. deep and held together by a simple locking device enabling them to be built up to reach any desirable height so long as they are prevented from being blown over. A garden or sun room suits them well, or they may be placed against a sunny courtyard wall at 5–6 ft high, secured by a strap placed around the 'tower', and fastened to the wall. Strawberries can even be grown in ordinary pots placed in a row at the foot of a sunny wall. All that is required is to keep an eye on their watering. A dozen good-sized pots will give several worthwhile pickings. To continue the fruiting, use another dozen pots in front of the first row and plant a later variety, possibly one which is autumn-fruiting.

In France it is quite common to find walls of a courtyard covered with fruit—peach and pear trees, and dessert gooseberries in the cordon-form, bearing large crops and ripening their fruit to perfection. Though this may not be possible in the climate of Britain, blackberries, and the other hybrid berries of similar habit, may be grown and trained to cover walls. They will fruit abundantly, even on a north wall, and will be attractive when in leaf and blossom. In the past, blackberries have been neglected because of their many thorns, but there are now thornless varieties—the new Oregon Thornless being worthy of planting for its handsome foliage alone, the leaves being deeply cut and fern-like. It will hide an unsightly brick wall, 6–7 ft high, in two years and give many pounds of fruit in autumn. If soil is absent from the courtyard, blackberries can be grown in large pots; this will enhance the appearance of the court-yard, especially if the walls are first treated with Snowcem. The long whipping shoots of the black-berries are tied in along strong wires which are held in place by threading them through nails made with an 'eye'. As blackberries mostly fruit on the old and the new wood, pruning requirements are at a minimum—just cut away any dead wood.

The Japanese Wineberry is another handsome wall plant; its stems are covered in crimson hairs and its fruit is delicious in tarts and flans.

Though not so accommodating as either the

strawberry or the blackberry, a whole new range of raspberries, introduced from East Malling by Mr Norman Grubb thirty years ago, have completely changed the outlook for the raspberry grower, extending the season by at least a month and providing him (or her) with frost-resistant varieties which produce almost double the amount of fruit from each cane compared with the weight obtained from older varieties. The canes are sturdy and grow up to 6 ft tall. Mr Grubb's first introduction, Malling Jewel, is the most reliable of all raspberries and also the earliest, but although it is first to ripen, it blooms quite late and usually misses the frosts. Malling Promise has been equally popular for although not quite so frost-hardy, it is virus-resistant and is the heaviest cropper of all whilst almost as early.

At East Malling, Dr Elizabeth Keep continues the work of her predecessor, Norman Grubb, in the task of finding a high-yielding disease-resistant raspberry which has, in addition, the ideal qualities for freezing, i.e. a berry that retains its moisture, does not go mushy when it thaws, and keeps its fullness of flavour. The new Malling Admiral, released in 1976 and having Malling Promise for a parent, appears, after exhaustive tests, to be the answer—the one most suitable for freezing. It is late to mature, thus extending the season and missing late frosts too.

Fruit for colder gardens
Few of those who like gooseberries, the first of the soft fruits after the all-too-often despised rhubarb, realise what a delicious fruit this is when grown for dessert. Few dessert varieties are now grown commercially, for the market grower concentrates almost entirely on the canner's favourite, Careless, which crops heavily and is bought by the housewife for culinary purposes. Only Leveller, which grows well only in parts of

Sussex and Hampshire and is the Cox's Orange Pippin of the gooseberry world, temperamental but delicious, is now grown commercially for dessert. It is left to the home grower to seek out those delicious old varieties and to grow them in the garden. If space is limited, gooseberries can be grown as single or double cordons against a sunless wall or along a garden path, like cordon apples. There are a number of varieties which make small compact plants and take up little space in the garden. Strawberries may be grown between the bushes for they enjoy the partial shade cast by the gooseberries when in leaf.

Those who know Langley Gage, which makes a small plant of neat, upright habit, and have tasted its white transparent fruit, moist with the mid-summer dew, will have tasted nectar, so superb is the flavour. Equally pleasant is Broom Girl; its golden-yellow fruits, shaded with green, being treacle-sweet when ripe. White Lion, almost the equal of Leveller in size, also makes good eating.

The gooseberry (and the blackberry) is the ideal fruit for those gardens situated north of the Trent. It prefers cool conditions (except for one or two varieties) and crops well and surely up to 1000 ft above sea level. Nowhere does the gooseberry grow better than in the Pennine regions of Lancashire and Yorkshire. It flourishes too as far south as Derbyshire and Staffordshire, and also in Cheshire. Until quite recent times, the old gooseberry shows of these regions (and that at Egton Bridge still survives), held in July, were the pride and joy of the local population—more popular than Bingo is today. At the shows, monster berries, as large as golf balls and weighing several ounces, were weighed at a certain time on a given day with all the ritual that goes with England's great occasions. The prize for the winner was usually a copper kettle!

For warmer gardens

'Gooseberries for the north, blackcurrants for the south' so runs the old adage. Blackcurrants require a warm garden, one sheltered from cold winds, and a heavy soil, whereas gooseberries prefer a light loam. Both fruits require exactly opposite conditions as to soil and climate. And whereas gooseberries are grown on a 'leg' and so do well as cordons, blackcurrants produce new growth from buds below and just above soil level in the form of suckers or shoots. They mostly make big bushes and need a great deal more room than many varieties of gooseberries.

Blackcurrants were mostly left to market growers catering for jam makers; they were also grown for their juice, which is rich in vitamin C. Then Laxton's Giant made its impact on the soft-fruit world. This variety bears fruit the size of an Early Rivers cherry and quite as juicy. It was the first real dessert blackcurrant and is sold in quarter-pound punnets at seaside resorts for visitors to eat as they walk about the promenades, but it is equally delicious in tarts and flans. It needs good culture, but a few bushes should be in every garden for it makes a plant of compact habit and is one of the first to ripen its fruit.

Fruit all the year

Beginning with early rhubarb, forced under a tea chest or a large upturned pot (and there is nothing more welcome in spring than the long pink sticks when cooked in a little brown sugar and served with shortcake), one may have a succession of soft fruits from early April until the year end when the blackberries and autumn strawberries complete the season. For at least nine months, and with the judicious use of cloches (which may be of glass or heavy polythene), one may obtain the best from the garden. What is more, soft fruits are all easily cultured, being more manageable than vegetables and tree fruits. There is nothing like as much to know about pruning, as there is with top fruits, or about the many forms in which they grow.

Once planted, soft fruits will almost perpetuate themselves. Rhubarb can be lifted and divided every fourth year. Gooseberries, correctly pruned, will remain healthy and productive for several decades (I know of gardens which boast gooseberries of fifty years old). Raspberries produce new canes each year and the old ones are cut out and destroyed; loganberries, and blackberries in most instances, bear fruit both on the old and new wood. Strawberries are propagated by removing the 'runners' which are attached to the main plants by 'strings'. The new plants will have rooted when cut away and re-planted in autumn. So, in most cases, the original cost of soft fruits is the only one. With care, the plants will last a lifetime and in the following chapters I will try to show how this is done and how to get the plants to produce the heaviest crops possible in the climate and soil of one's garden.

Soft fruits come quickly into bearing—strawberries within eight to nine months of their planting; rhubarb, blackberries and gooseberries within a year or so. There is no long waiting as with tree fruits—sweet cherries taking ten years to bear a worthwhile crop.

The following is a monthly table of soft fruits:

JANUARY
Rhubarb (forced in a shed or garage)

FEBRUARY
Rhubarb (forced)

MARCH
Rhubarb (forced in the open)

APRIL
Rhubarb
Strawberry: Cambridge Brilliant
(under cloches)
Cambridge Premier
(under cloches)

MAY

Gooseberry: May Duke
Strawberry: Cambridge Brilliant
and Premier

JUNE

Gooseberry: Broom Girl
Keepsake
Whitesmith
Strawberry: Cambridge Early Pine
Cambridge Favourite
Red Gauntlet
Royal Sovereign

JULY

Blackberry: Bedford Giant
Blackcurrant: Laxton's Giant
Mendip Cross
Wellington XXX
Gooseberry: Cousen's Seedling
Gunner
Leveller
Raspberry: Lloyd George
Malling Jewel
Malling Promise
Strawberry: Cambridge Late Pine
Fenland Wonder
Gento
Talisman

AUGUST

Blackberry: Bedford Giant
Merton Early
Blackcurrant: Seabrook's Black
Westwick Triumph
Gooseberry: Drill
Howard's Lancer
Leveller
Loganberry
Raspberry: Glen Clova
Malling Enterprise
Norfolk Giant
Red Currant: Laxton's No. 1.
Strawberry: Cambridge Rearguard
Gento
Hampshire Maid

SEPTEMBER

Blackberry: Himalaya Giant
Oregon Thornless
Blackcurrant: Amos Black
Daniel's September
Blueberry
Raspberry: Malling Landmark
September
Red Currant: Raby Castle
Red Lake
Strawberry: Hampshire Maid
La Sans Rivale

OCTOBER

Blackberry: Himalaya Giant
Oregon Thornless
Blueberry
Hybrid Lowberry
berries: Japanese Wineberry
Raspberry: September
November Abundance
Strawberry: Hampshire Maid
La Sans Rivale
St. Claude

NOVEMBER

Strawberry: La Sans Rivale
(under cloches)
St. Claude
Gento

DECEMBER

As for November

2 Starting a Fruit Garden

Use of fencing

Apart from rhubarb, which will grow almost anywhere, and blackberries, which may be grown on a north wall, most soft fruits require some sunshine to ripen the fruit. Gooseberries, however, grow quite well in semi-shade being naturally plants of the hedgerow. Many suburban houses have a long narrow garden with a hedge or fence to divide it from the neighbour's garden. A fence, usually made of interwoven or overlapping timber, could be enhanced by planting blackberries, one plant to each 6 ft panel, and training them along wires fastened to the panel frames. Fencing which is more shaded could be used for blackberries, and that which receives a greater amount of sunshine could be used for raspberries. The raspberries should be planted 15 in. apart and should be cut down in their first spring to 3 in. above soil level. New canes will arise from the base buds to bear fruit the following year. The new canes should be tied to wires stretched across the fencing panels at 2 ft and 4 ft from the ground. Strong galvanised wire will last many years, and so will the raspberries. At least a year before planting anything against them, the panels of the fence should be treated with creosote to preserve them. If erecting a fence, make sure to use posts of 4 in. × 4 in. timber and cement them in, for high winds will play havoc with insecure fencing.

Making a fruit garden

Perhaps the end of the garden would be the most suitable place for making a fruit and vegetable plot. It need not take up more than the area of a generously sized sitting room, say, about 24 ft × 20 ft; a plot of about 28 ft × 42 ft would be ideal. An attractive screen of rustic poles could divide the plot from the rest of the garden, i.e. from the lawn and flower beds, though a well-kept fruit garden has an orderly appearance. An open situation is necessary and planting too near tall trees should be avoided.

If there is an interwoven fence at the end of the garden, it can be used for growing blackberries against; or they can be grown against a rustic screen, where the fern-like leaves of Oregon Thornless are most handsome if the long shoots are tied in as they form.

At one side of the fruit garden, which is entered, perhaps, through a rustic archway at the centre or to one side of the screen, several rows of raspberries should be planted. At the end of each row strong stakes, 5–6 ft high, should be inserted. These will carry galvanised wire which is stretched along the rows and to which the new canes are fastened to prevent their being broken by strong winds. The wires should be in four lengths at intervals of about 15–16 in. apart. If planted 12–15 in. apart, about eighteen canes would be required for a 20 ft row. Three varieties should be grown to fruit early, mid-season and late, perhaps deciding upon Malling Jewel (or Malling Promise), Glen Clova and Norfolk Giant. Make the rows 2 ft apart. Though they produce new canes each year on which they fruit, the raspberries will be permanent and so should be away from general cultivations.

Then plant two rows, 4 ft apart (and the same distance apart in the rows), of blackcurrants, five plants to each 20 ft row and in perhaps two varieties. One should be the early-fruiting Laxton's Giant and the other the ever-reliable Wellington XXX. If space is strictly limited, plant instead Amos Black or Westwick Choice, though the more vigorous varieties may be controlled by pruning, removing some of the older wood as it forms. Like raspberries, black-currants should be cut back to within 3–4 in. of the base in March after planting. This is done to stimulate the formation of new shoots, so there will be no fruit the first summer.

However, the ground need not remain unpro-ductive during that time if strawberry runners are planted between the rows. Plant them as a double row 12 in. apart and 9 in. in the rows. They may remain undisturbed for at least two years or until such time as the currants will have formed large bushes. If the strawberries are planted in late autumn, at the same time as the currants, they will bear quite a good crop the following summer, eight to nine months after planting. Plant an early variety and cover with cloches in April, or plant an autumn-fruiting kind.

Also in your soft fruit garden should be a row of gooseberries. Once established, gooseberries crop so heavily that two or three plants of each of three varieties will give sufficient pickings at one time to make a large pie or flan, or provide dessert for a family of four. Spread the season by planting at least three kinds; Keepsake is the first to mature, usually cropping heavily; Broom Girl is another early fruiter. To follow comes Whitesmith and the later Golden Gem, reliable in most soils and situations. These possess outstanding flavour but the true dessert varieties, the monsters, which in the hands of the experts reach golf-ball size and have treacle-like juice, each variety with its own particular flavour, should be grown as double or single cordons and nothing could be easier. From one-year plants you can form your own but they need to be grown in rows and trained against wires like cordon apples. See Chapter 4.

If the garden is exposed and open to cold winds, omit the currants and instead plant more gooseberries. Here again, strawberries can be grown between the rows, which will be 3–4 ft apart. Plant a double row and if an early variety, plant so that the rows can be covered with cloches to ripen the fruit by the end of May. Then follow with a mid-season variety and here, Royal Sovereign is supreme for flavour. Alongside, allowing 12 in. between the double rows, plant the later-maturing Talisman or Hampshire Maid. With strawberries it is important to spread the season so that if some plants (those that bloom first) are caught by frost, others will escape. Though they will fruit later they will be none the less welcome. By planting a double row of four or five kinds, which will take up little space, much of the ground will be productive almost from the beginning and one can never have too many strawberries, though they are more appreciated if the supply is regulated throughout summer and autumn.

A few plants of red currants will yield a good crop of this delicious fruit though some precaution should be taken against birds to which the bright red berries are always an attraction. Red Lake is the best variety. Of compact habit, it may be planted 2 ft apart each way. Like gooseberries, red currants grow on a 'leg' and so may also be grown as a single or double cordon.

Then plant a row of rhubarb, perhaps ten roots spaced about 2 ft apart; say, two roots of five varieties—early, mid-season and late. The first to mature is Timperley Early and if the roots are covered in early February, there will be succulent pink sticks 2 ft long to pull during March. At this time of year, stewed rhubarb can taste like peaches in summer, and if there is too much to use at this time, put some in the deep freezer, for no other fruit behaves better when frozen. Uncovered roots of Timperley Early will be ready when those which were covered have finished. After removing the covers, do not pull any more sticks during that year. The uncovered roots will give a long succession of succulent sticks until the end of summer, though when other fruits are available, rhubarb takes a back seat. It is always most in demand in spring and early summer.

To complete the soft-fruit garden, plant a row of loganberries or of other hybrid berries. They will need training along wires and loganberries are suitable only for wind-free gardens. In colder parts grow blackberries instead or several of the hybrid berries which are hardier.

The soft fruits will use up about half the area allocated for fruit and vegetables, but by planting against fencing and growing some fruits as cordons, together with inter-planting, full use of the available space will be made and every inch of ground is now allowed to earn its keep.

The following selection of fruits should prove capable of yielding large crops over a long period:

	Gooseberries	*Strawberries*
Early	Keepsake	Cambridge Favourite
	May Duke	
Mid-season	Whitesmith	Royal Sovereign
Late	Leveller	Talisman
	Raspberries	*Blackcurrants*
Early	Malling Jewel	Laxton's Giant
Mid-season	Glen Clova	Wellington XXX
Late	Norfolk Giant	Amos Black

Frost-hardy varieties

For low-lying, frost-troubled gardens, perhaps those situated by the banks of a river, fruits which either bloom late or are frost resistant should be grown. Gooseberries will rarely be harmed by frost although,

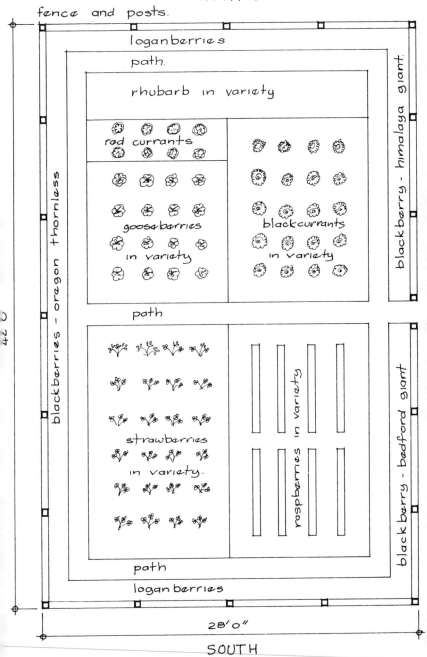

Fig. 1 An ideal layout for a soft-fruit garden.

to play safe, it is advisable to plant the later-maturing varieties which also tend to bloom later. In contrast, blackcurrants are often severely damaged by frost, so where frost is troublesome, plant the resistant Seabrook's Black, Wellington XXX or Mendip Cross, the last named being the most highly resistant of all. Amos Black is also resistant, whilst Baldwin and Westwick Choice are the most prone to frost damage.

In raspberries, Malling Jewel blooms later than other earlies and is the best for frosty land. Its sister, the later-maturing Malling Enterprise, is equally resistant and should be the choice for a mid-season variety. To complete the planting, the later-ripening Norfolk Giant blooms so late that it misses late frosts everywhere.

Strawberries, but not the autumn-fruiting kinds like La Sans Rivale, are even more liable to be spoilt by frosts than raspberries for they bloom in May, often a difficult month for soft-fruit growers. One can either plant those which bloom late, such as Talisman, Fenland Wonder or Cambridge Queen, or those which have shown a marked resistance to frost, such as Cambridge Regent and Early Pine. Here is a selection for frost-troubled gardens:

Gooseberries	*Strawberries*
Dan's Mistake	Cambridge Regent
Howard's Lancer	Fenland Wonder
Gunner	Talisman
Raspberries	*Blackcurrants*
Malling Jewel	Mendip Cross
Malling Enterprise	Seabrook's Black
Norfolk Giant	Amos Black

Soils and their preparation

Fortunate are those whose garden soil is a well-drained loam. Usually, when moving to a new house it will be found that the top soil has been removed and much needs to be done to bring the ground into a condition in which soft fruits will grow well. If the soil is heavy, containing a too large proportion of clay causing it to be waterlogged in winter and to bake hard in summer, much will need to be done before planting. Such a soil will rarely grow good strawberries for in a badly drained soil the plants will suffer from Red Core root rot for which there is no cure. Only perhaps Huxley's Giant and Talisman, both highly resistant to this trouble, can be expected to produce a reasonable crop. Blackberry, Himalaya Giant, and blackcurrants, especially the vigorous Wellington XXX, will do well, as will the equally vigorous Whinham's Industry and Howard's Lancer in gooseberries, though this fruit always does better in light soil. Raspberries, too, rarely give a good account of themselves in heavy, badly drained soils, though the robust Malling Promise may crop reasonably well.

Soft fruits may be planted at any time between October (for strawberries) and March. The soil should be prepared during the winter months but preferably between October and the year end for 'as the days begin to lengthen, the cold strengthens' runs the old adage, and between January and early March the soil is often too frost-bound to work. There is then little time left for the planting programme to be completed. Again, if planting can be done in November, when the soil is still warm, the plants will become established and will have begun to form new roots before the hard frosts set in.

If the soil contains a higher proportion of clay than permits it to be readily worked, cover the surface with caustic (unhydrated) lime obtained from a builder's merchant and dig it roughly in. When the lime comes into contact with moisture, a violent reaction takes place causing the clay particles in the soil to break up and the lime to disintegrate. This will also serve another purpose in that the soil will be limed at the same time. Of the soft fruits, only blueberries and strawberries prefer a slightly acid soil and this, only when the soil is in good heart. Apart from its ability to correct acidity, lime is able to release the plant foods pent up in the soil so that even well-manured land will not prove beneficial to the plants where lime is not present to unlock the foods. Without lime to open up the clay particles, a heavy soil will so consolidate in winter as to deprive the plants' roots of the necessary oxygen for their active growth; also, without oxygen, bacteria in the soil will be unable to fulfil its job of converting humus into plant food. As a general rule, about 14 lb of hydrated (or unhydrated on heavy land) lime should be applied to every 200 sq ft of ground.

Aerosil, a product of British Gypsum Ltd, will help to bring a heavy clay soil into a more easily worked condition, whilst it will bind together the particles in a sandy soil and so improve moisture retention. It also provides valuable mineral salts, such as magnesium and calcium.

Several compounds, such as Krillium and Colimnus, if applied when the soil is reasonably dry in September, will help to bring the soil into better condition. Just sprinkle the conditioner over the surface and dig it in. A 14 lb bag will treat 200 sq ft of ground. These compounds are expensive and should be treated with respect.

A clay soil will consist of about one-third clay and, in addition to the lime, as much drainage materials and humus as can be obtained should be worked in to open it up. Shingle, crushed brick (often present on a building site) and clearings from ditches should be dug in and any form of humus obtainable. Decayed leaves and garden compost are suitable, also peat which should be augmented by farmyard manure. Those situated in the north will be able to obtain quantities of cotton or wool shoddy (waste), which is inexpensive, whilst used hops from breweries are often to be had for the taking. Those living near the coast will be able to obtain seaweed, but before digging in, it should be chopped up with a sharp spade to enable it to break down more quickly.

Where none of these supplies of humus are obtainable, a quantity of straw (a bale) may be composted. Erect corrugated iron sheets on three sides and made about 5 ft square by driving stout stakes into the ground, one on either side of each iron sheet; or use old weather boarding. Spread out the straw, which has first been made moist, and sprinkle over each 12 in. layer of straw, an activator or compost maker, such as Adco or Garrotta, and water it in. Proceed in this way until the pile is built to a height of 4–5 ft. Cover the tip with sacking and leave it a week or so to heat up. Remove two sides of the sheeting and re-erect them to form another square, leaving the remaining sheet in place for one side, then turn the heap, shaking out the straw and giving more water if it appears dry. After another week it will have begun to turn rich brown and be generating considerable heat. After one more turn, it will be ready to dig into the soil. It may be augmented by some peat or other material, such as lawn mowings and decayed 'greens' from the compost heap. It will heat more readily if some poultry manure is added.

Light land, containing about two-thirds sand, should be treated in the same way. It requires humus to retain moisture in summer whereas heavy land needs it to open up the soil. But lucky is the gardener whose soil is light, for it is easier to manage and may be worked at all times of the year except when frosted.

All soft fruits, especially gooseberries and straw-berries, do well in light soils but these are hungry soils and require continuous supplies of humus and plant food to maintain fertility, for in such soils, plant food is readily washed away.

Peat is a valuable source of humus for all soils and the sphagnum moss peats, which are only partly decomposed, are able to hold twenty times their weight of moisture, thus being indispensable for light, sandy soils. A bale of peat moss (14 bushels) will cover an area of 200 sq ft to a depth of 1 in. It can be applied before the manures and other forms of humus, to be dug in together. Peat is of value for mulching strawberries, raspberries and gooseberries and can be given right up to the plants. It will conserve moisture in summer and suppress weeds as well as supplying the soil with a constant supply of humus. It will also do away with the need to hoe near the plants, which could cause root damage.

Soils of a calcareous nature are usually 'hot' soils. They have only a thin layer of top soil and in summer become so hot and dry that they are unable to support plant life unless they are given copious amounts of moisture. For soft fruits to be successful, it will be necessary to increase the depth of top soil and this may be done by working in all forms of humus, especially peat. 'Green' manuring will have consider-able value.

This is done by covering the surface, having cleared it of weeds, with rape seed. The seed should be sown thickly and raked in, August being a suitable time to do so. When it has grown 2–3 in. tall and by which time it will have formed a thick mat of fibrous roots (this will be about the end of September), dig it all in as deeply as possible. It will decay during winter and will add both humus and plant food to the soil. Other humus materials can then be added too.

Food requirements

Soft-fruit plants need a balanced diet: nitrogen to promote a continuous supply of healthy new wood and to increase the size of the fruits; phosphates to bring the crops to maturity and to promote healthy root activity; and potash to build up a 'hard' plant, one able to withstand adverse weather. Potash will also improve the flavour, colour and quality of fruit and prevent it from going soft too soon. Gooseberries, which grow best in a light soil, require more potash than, say, blackcurrants, which crop best in a heavy soil, and because gooseberries bear fruit both on the old and new wood, they do not require such large amounts of nitrogen as blackcurrants which fruit best on the new wood. This they are required to form continuously and in as large a quantity as possible. The more new wood they make, the heavier the crop.

Farmyard manure and old mushroom compost contain all the necessary plant foods, as well as valuable trace elements, and release them over a long period. Seaweed, shoddy and used hops are rich in nitrogen, likewise fish meal and dried blood. Fish meal and guano are quick acting and both have a high potash and phosphate content. Poultry manure is also rich in all the plant foods. These are known as organic manures, being natural foods, and as well as supplying the plants with food they also provide valuable humus which the inorganic fertilisers do not. Those, like nitrate of soda and sulphate of potash are chemically produced. They are useful in that they are quick acting. Nitrate of soda may be used sparingly in spring to start strawberries and blackcurrants into new growth whilst all soft fruit will benefit from a 1 oz per yard dressing of sulphate of potash to increase the quality of the fruit. But they are best used in conjunction with organic fertilisers. Sulphate of potash can be replaced by wood ash (bonfire ash) which has a reasonable potash content and which should be stored dry, otherwise the potash will be washed away. Soil containing nitrogen but deficient in potash will produce lush, soft plants, prone to mildew and botrytis and which will fall to the first frost and cold winds.

The following is a list of valuable fertilisers used in growing soft fruits:

Fertiliser	Action	Nitrogen content (percentage)	Phosphate content (percentage)	Potash content (percentage)
Basic slag	Slow	15	—	—
Bone meal	Slow	5	20	—
Dried blood	Medium	10	—	—
Farmyard manure	Slow	0.5	0.25	0.5
Fish meal	Quick	10	8	7
Guano	Quick	15	10	7
Nitrate of soda	Quick	10	—	—
Nitro-chalk	Quick	10	—	—
Poultry manure	Medium	3	2	6
Rape meal	Slow	5	2	1
Seaweed	Slow	5	—	1.5
Shoddy (wool)	Slow	12	—	—
Sulphate of ammonia	Quick	20	—	—
Sulphate of potash	Quick	—	—	50
Superphosphate	Medium	—	15	—
Used hops	Slow	4	2	—

Working the ground

Having obtained the necessary fertilisers and humus materials, it is now important to incorporate them into the soil. This should be done to as great a depth as possible as most soft fruits are deep rooting. At the same time, the soil must be brought into a friable condition. In consolidated ground, soft fruits will be a failure because excess moisture cannot drain away and the roots will decay. Also, wet, badly drained soil will not heat up with the early spring sunshine and much time will be lost in bringing the much appreciated early crops to maturity.

It is advisable to trench the soil at least to the depth of a spade, incorporating the humus materials to something like 9 in. deep. This will keep the roots moist well below the surface, for if the roots dry out, the fruit will be hard and seedy, entirely devoid of juice and flavour.

To trench the ground, first remove the soil at the top of the piece to be prepared and place it at the other end. This will be used to fill in the last trench. Then, into the trench, at the top, add drainage materials, if necessary, and manures. Incorporate the humus with the soil which is being turned into the first trench and continue in this way until the whole piece has been treated and the available humus has been used. It is a good idea to cover the surface with peat and some organic manure (the rest being placed at the bottom of the trench) before digging begins, for in this way it will be better distributed. Autumn is the best time to do the work and if the surface is left in a rough condition for the frosts to pulverise, the ground will be in a fine tilth when planting is done. With heavy ground, the work is best left until March, but where the soil is light, it will already be in a suitable state for planting. This is done after allowing the soil a few days to settle down.

As the digging continues, remove all perennial weeds, such as nettles and couch grass, and dig in the annual weeds. But do not dig the soil if it is saturated by heavy rains or if there is frost in it. Wait until it is reasonably dry and the digging will be so much easier.

Mulching is also important. This means covering the soil around the plants with some form of humus materials. It should be done early in summer for it will suppress weeds and help the soil to retain its moisture by preventing rapid evaporation. It will also provide the plants with a boost to their diet by its additional food value. It should be gradually worked into the soil during cultivation and so increase the fertility. Decayed farmyard manure and old mushroom-bed compost are suitable for mulching and these can be augmented with some peat, which on its own will act as a useful mulch, especially for strawberries which enjoy its slight acidity. As it is quite inexpensive, it can be applied to a depth of about 1 in. It will prevent the soil from 'panning' as so often happens when heavy rain is followed by hot sunshine. Used hops and well-composted straw are other suitable materials for mulching.

A few tools will be essential to do the work well. In addition to a spade, a border fork, a hoe and a good trowel, of stainless steel if possible, will be needed and a garden line for planting in rows. One should also have a stick or cane 5 ft long and marked in feet so that the plants may be set out at the correct distances. A watering-can or hose fitted with a sprinkler will also be needed for in a dry summer, regular watering will be necessary to swell the fruit and keep it juicy. Hard, seedy fruit, due to lack of moisture, is not required. An old tub or barrel for manure water (see page 35) will also enhance the quality of fruit by its use.

3 Rhubarb

General considerations
Though really a vegetable, rhubarb is always grown as a fruit and used for the 'sweet' course, served with custard or cream after stewing or making into a pie. And what could be more welcome early in the year than those succulent pale-pink sticks removed from

three-year roots which have been lightly forced in a shed or garage, perhaps in a little gentle heat?

Taking up only limited space in the garden, one established root occupying about a square yard, it is a hardy and economical fruit, yielding twenty to thirty sticks during the year and it is almost indestructible. It suffers from neither pests nor disease, is tolerant of a heavy soil and no amount of frost will harm it. Indeed, it forces better after exposure to frost. Those who enjoy its flavour rarely tire of it all the year round. But it is early in the year, when there is no other soft fruit about, that the demand for rhubarb is at a peak and this lasts until the first of the outdoor strawberries are ready early in June. This is the time when rhubarb is at its best and when any surplus should be placed in the freezer.

Rhubarb requires a heavy soil for it to make those thick juicy sticks and as much humus as possible must be dug in before planting. The centre of the rhubarb industry is mid-Yorkshire where it receives ample supplies of wool and cotton shoddy which suits it well. But farmyard manure, garden compost and old mushroom manure are equally good and each of these manures may also be given as a top dressing when the roots are cleared of their dead foliage in autumn. In this way, the roots, or crowns as they are called in the trade, will retain their vigour for many years. At planting time, give a 4 oz per square yard dressing of basic slag, which releases its nitrogen over a long period, and it is nitrogen that rhubarb loves best.

Growing from roots

Rhubarb roots, or thongs, which can be obtained from specialist growers, contain an 'eye' or bud from which the sticks grow. The leaves, small at first, unfold from the buds before the stick appears. The roots may be planted any time from mid-October until January when they are dormant. Plant 3 ft apart, with the soil just covering the bud, and tread firmly round taking care not to damage the bud.

After about four years, the roots will have grown as large across the top as will just about fit into a good-sized bucket. They should then be lifted and divided, perhaps one root each year so that a generous supply of sticks will be maintained. Lift with a strong fork and use a sharp knife to make the divisions. Each should have an 'eye' or bud and a piece of root attached. Treat the cut part with hydrated lime and plant as soon as possible. It may be that at the centre of the root there is a small area which has begun to decay. If so, this must be discarded. In this way, the stock will be kept in a healthy condition for years.

Growing from seed

Rhubarb plants may also be raised from seed with a minimum of trouble and expense. Obtain a reliable strain and sow in drills 9 in. apart and 1 in. deep in early April. The seed bed must be in a friable condition and before sowing, 4 oz per square yard of hydrated lime should be raked in. Older roots should also be given a lime dressing in winter as this will prevent the sticks becoming soft during wet weather.

Sow the seed thinly so that the young plants will not need transplanting before moving them to their permanent quarters. Keep them moist and regularly hoe between the rows. By early autumn, the young plants may be set out 18 in. apart where they are to fruit. The following year, pull only one or two sticks from each root, leaving a number to die back in autumn and fortify the roots. Next year, more sticks can be removed and after three years, lift each alternate root and replant 3 ft apart, for by then they will have grown quite large. After a further year, lift and divide the other roots.

Obtaining an early crop

To obtain early sticks, lift two or three roots of three- to four-year plants about mid-December and leave them on the ground to become frosted. They respond to this shock treatment by coming into production more quickly when taken indoors.

The roots may be planted early January in a deep, wooden orange-box, with friable loam and some peat packed tightly between the crowns, which may be almost touching each other. Water the roots and place the boxes in a cellar, shed or garage, preferably in the dark, or cover the top of the box with sacking. Alternatively, the roots may be placed on a 6 in. layer of soil and peat wherever they are to be forced, packing them closely together with soil around them. The tops of the roots, the 'eyes', should be just showing. In the warmth of a building, the sticks will soon appear and by early February (earlier if the place is heated) the first will be ready to pull. Always pull the sticks when about 18 in. long by holding them firmly near the base and they will not break. Before cooking, cut off the leaves at the top. Make sure that the soil about the roots is kept comfortably moist but only small amounts of water will be needed at this time, unless growing in heat.

Roots may also be forced in a cold or heated frame, following the same procedure and covering the frame lights with sacking to exclude most of the light.

Again, roots may be forced where they are growing, when the sticks will be ready about a month after covering. The plants must be three to four years old before they can be forced in this way. Do not lift the roots; simply cover them where they are. Early in February, cover the roots with some farmyard manure and over them place a deep box, barrel or tub, or even a bucket, resting on stones to admit air. Another method is to erect a deep frame around several roots, using corrugated sheeting for the sides or deep boards held in place by strong stakes driven into the ground. Over the top place several boards to which sacking is nailed to exclude light. The roots should be given a 6 in. layer of manure or composted straw before covering and should be kept nicely moist. By early March the first sticks will be ready to pull and may be used when 12 in. or more in length. They will grow straight and be tender when cooked.

Extra early sticks can be obtained by planting the roots in a hot bed, surrounded (as described) by corrugated sheeting. Lift the roots in December and expose them to frost. At the same time begin preparing a hot bed by composting straw with an activator and to which a quantity of fresh farmyard or poultry manure has been added to enable it to generate more heat. It will take three weeks to prepare and by then will have turned rich brown and will have reached a temperature of 100 °F at the centre of the heap.

Early in January, remove soil to a depth of 12 in. and bed down the compost, treading it firmly to retain the heat. Then cover it with 6–8 in. of soil (previously removed) and peat into which the roots are planted. Cover with sacking and keep moist. By the end of January the first sticks will be ready to pull.

Fig. 2 Forcing rhubarb in the open with the aid of a flower pot.

compost

Forced roots will bear sticks for about two months. Those which have been lifted should be re-planted into soil freshly fortified with manure. They should have no further sticks removed that year and other roots should be forced next winter. It will take the forced roots eighteen months to recover their vitality, but if about eight to nine roots are grown, this will allow several to be forced each year in succession and still maintain their vigour.

Growing in the open

Those roots which are not in any way forced but are growing naturally will begin to form long stems by early April and will bear many succulent red sticks before the strawberries and gooseberries are ready early in June. Freeze the sticks at this time, before they grow coarse and stringy. By mid-July they will have passed their best and should be left to die down gradually. The stems and leaves are then removed and burnt and the roots given a liberal top dressing of decayed manure.

Towards the end of August, some roots will form seed heads. These should be removed as soon as noticed, otherwise the energies of the plant will be directed to the production of unwanted seeds. At this time, the plants will benefit from an occasional application of diluted manure water. This is made by filling a sack with farmyard or poultry manure and immersing it in a tank or barrel of water for a week or more. The sack should then be removed and the tank filled up to dilute the manure water. This nutritious liquid can also be used for strawberries and gooseberries, greatly increasing the size of the fruits and their flavour. Use manure water when the soil is damp, or water the ground after its use so that it will reach the roots of the plant before it can evaporate. It should be used once a week during the summer months.

Deep freezing

Rhubarb is excellent for freezing and will retain its freshness for at least eighteen months. Use the sticks when red and tender, before the end of June. After pulling the sticks, cut them into pieces 1 in. long, place in plastic bags and consign to the freezer. About three hours should be allowed for it to thaw before cooking. Rhubarb leaves should not be used for any purpose.

Varieties

EARLY

Hawke's Champagne. One of the earliest for forcing, producing strong, thick sticks of deep crimson.

Royal Albert (also *Early Albert*). The most popular early variety for more than a century. It forces well and also comes early outdoors, producing large stems of bright scarlet and of excellent flavour.

Timperley Early. A new early of splendid quality, being hardy and bearing stout sticks of deepest crimson which are sweet and juicy.

MID-SEASON

Canada Red. Excellent to force outdoors, it produces plenty of thick, crimson sticks over a long time.

Cawood Castle. A new variety raised at the Stocksbridge Experimental Station and bearing an abundance of long, red sticks of excellent flavour.

Dawe's Challenge. A mid-season rhubarb which does well in all soils and in all climates, the large handsome sticks having good flavour.

Stott's Monarch. Unique in that the stems mature a deep green colour and when cooked have a delicious pine flavour. Use it to mix with other varieties when cooking. A strong-growing variety, it should be given more space in the rows.

Glaskin's Perpetual. Grows well from seed and produces its large green sticks, shaded red over many weeks, making delicious eating even towards the end of summer.

Myatt's Victoria. The last to mature and with its handsome large red sticks of excellent flavour, it is the canner's favourite.

The Sutton. Late, it bears its vivid red sticks over a long period for it does not set seed and devotes its energies on its succulent stems.

4 Gooseberries

Where to grow them

The gooseberry, with the blackberry, is the hardiest of the soft fruits. *Ribes grossularia* is a native plant, to be found about old ruins and rocky outcrops usually high above sea level. It also grows in woodlands, and in the wild produces small red fruits. Mentioned by Tusser in 1580, it was esteemed by country folk as it was the first fruit of the year, so welcome after a long winter when the diet was sparse. It can be made into pies or bottled to use in winter, whilst it makes an excellent jam or jelly; also a delicious gooseberry vinegar. In cold, exposed gardens, where most fruits fail to do well, gooseberries can be relied upon to yield good crops in most years. By planting early, mid-season and late kinds, fruit may be expected from the end of May until September, when blackberries and late (autumn) fruiting strawberries are ripe and will continue until the year end. Gooseberries are troubled neither by birds taking the fruits nor by frost or cold winds.

At one time, and especially in the north, dessert gooseberries were more widely grown than strawberries, which fruited all at the same time and if late frosts prevailed, might bear few fruits at all. Dessert gooseberries could always be relied on and where early, mid-season and late varieties were grown, there was choice fruit over a period of many weeks, from the end of May until early September. Well known is May Duke, which is the first to mature, and Careless, which comes later. Both are culinary varieties, cropping heavily and which freeze and bottle well. They should be grown in every garden but the delicious dessert varieties are comparatively unknown today and we miss much for not even strawberries make such good eating as dessert gooseberries grown well.

A cool climate enables the fruit to mature slowly. This brings out the subtle flavour of the berries and it is surprising how different each variety tastes. Gooseberries do well up to 1000 ft above sea level and no amount of cold wind troubles them. It is said that it is not fashionable to plant gooseberries for because of their thorns, picking is never enjoyable but against this, the berries weigh heavily and a pound of fruit may be picked more quickly than a pound of currants or raspberries. Also, this is one of the few soft fruits to hang on the plants for several weeks after ripening without their going 'mushy' and losing quality. Gooseberries can be picked when one has the time to spare. They will also freeze well, retaining their juice and flavour for at least eighteen months, and when thawed they are as if only just gathered.

Situation matters little for gooseberries. The more vigorous kinds, e.g. Howard's Lancer, may even be planted as a wind break against prevailing winds. Soil, however, plays a large part in obtaining a heavy crop. The plants enjoy light land, well drained in winter but which contains plenty of humus to hold summer moisture. Without this the plants will not bear heavily, neither will the fruits swell up and be juicy. In particular, ample supplies of moisture are necessary for dessert fruit. Plenty of humus will also help the soil to remain cool in summer for which reason gooseberries never crop well in shallow, chalky soils which become too hot and dry in summer.

Soil requirements

With gooseberries, it is important to maintain a balance between the old and the new wood for the fruit is borne on both. Excess nitrogen should not therefore be given, rather should the plants be grown 'hard'; nitrogen will encourage mildew to which gooseberries, in too rich a soil, are so liable. At the same time, sufficient nitrogen must be provided for the formation of some new wood each year otherwise the plants will die back from an excess of old wood.

Decayed farmyard manure, old mushroom compost, used hops and composted straw are suitable for providing humus. They are better for gooseberries than shoddy and poultry manure which are richer in nitrogen and better for blackcurrants. Dig the materials in as the ground is prepared, taking care to remove all weeds for established plants resent disturbance of the roots. As with most soft fruits, the roots are formed just below the soil surface and travel a distance from the plant, thus making cultivations difficult after planting. Rather than work the hoe too near the plants, it is better to rely on a mulch to suppress weeds.

Where these manures are unobtainable, dig in 2 oz of bone meal or steamed bone flour per plant or per square yard of ground and this will release its nitrogen content slowly over a long time. Bone meal is also rich in phosphate which will stimulate root action so that the plants will search far and wide for food and moisture which is greatly to the benefit of the crop.

It should be said that those varieties which tend to make little new wood, e.g. London, Lord Derby and Princess Royal, will need larger amounts of nitrogen to stimulate them into making wood. These varieties need little pruning.

Gooseberries are potash lovers, as are most plants which grow better in light soil for the potash content is readily washed away and so needs replenishing every year. Just before planting, or in March if planting takes place in autumn, rake into the surface 2 oz of sulphate of potash per square yard or per plant. As an alternative, use bonfire ash which has been collected and kept dry. Repeat the potash dressing every spring.

For a mulch, which is given early in May, use decayed manure or composted straw together with a quantity of peat; or the peat can be applied first and then the compost. See that it is placed right up to the leg of the plant and around it, to cover an area of about a square yard. A mulch will prevent moisture evaporating from the soil and will also do away with the need to hoe near the plants. The use of well-decayed manure will provide the plants with as much nitrogen as they require to form new wood.

Where growing dessert varieties, an occasional application of manure water will increase the size of the berries. Again, never allow them to lack moisture or they will not attain a good size. During hot, dry weather, give each plant a gallon of water every other day, preferably in the evening so that it can reach down to the lower roots before it evaporates. If the soil is dry, begin this extra watering as soon as the fruits have formed and are about pea size. If left until they start to ripen, the skins will burst and the berries will never reach a good size. More than any other crop, gooseberries will respond to careful feeding and watering during the early summer months and May is often the driest month of the year.

Planting

Gooseberries can be planted at any time from early November until early March, except when there is frost in the ground. In heavy land, March planting is preferable. Where growing in the bush form, plant those of spreading habit about 5 ft apart each way, and those which grow upright, only 4 ft apart. Those making spreading plants are: Careless, Keepsake, Howard's Lancer, Whinham's Industry, Whitesmith, Leveller, Lancashire Lad, Thatcher, Crown Bob and Early Sulphur. Those of neat, upright habit are: Laxton's Amber, May Duke, Bedford Red, Langley Gage, White Transparent, Green Walnut, Gunner, Drill, Golden Gem.

Where space is strictly limited, slightly closer planting of the upright varieties is permissible but here the plants must be kept under control by regularly cutting out dead and excess wood to prevent the plants growing into each other.

Plant firmly, treading the soil around the leg but taking care not to injure it. Spread out the roots and make the hole large enough to permit this. Gooseberries will not do well if the roots are put in the ground bunched together. The top or surface roots should be only just beneath soil level when covered. Before replacing the soil, it is advisable to cover the roots with peat if the soil is heavy.

Where growing culinary varieties, where large quantities of berries are required and size is immaterial, the bushes will be given the minimum of pruning and none for the first three to four years. Afterwards, remove any dead wood and cut back the shoots where there is overcrowding. Those of spreading and drooping habit, such as Keepsake and Whinham's, should have the established shoots cut back to an upward bud to counteract this drooping tendency. Those of upright habit are cut back to an outward pointing bud, to prevent overcrowding at the centre of the bush. Gooseberries vary greatly in habit (some bear most of their fruit at the centre, others at the outside) and each should be pruned accordingly.

Where growing for exhibition or top quality dessert fruit, the new wood is cut back each winter to about two-thirds. This will direct the energies of the plant to the fruit rather than to the formation of an extension to the shoots. With cordons, prune back or pinch out the shoots to within 3 in. of their base and this should be done in March. Of course, all dead wood should be removed each winter as routine.

Again, where top quality dessert fruit is required, a certain amount of thinning is needed in those years where there is a heavy set of fruit. But this should not be done until the berries have begun to swell for birds may have taken some and others may have dropped during cold winds. Thin only where there is over-crowding as the fruits swell and colour.

Propagation

Gooseberries are grown on a 'leg' to prevent the formation of suckers. On the other hand, black-currants rely on the continual appearance of suckers for their replacement wood. Gooseberry shoots (cuttings) are always more difficult to root for the wood is harder. Because of this, use only new shoots, the wood being of light greenish brown against the dark brownish grey of the older wood. September is the best time to take cuttings of the new season's wood and it is important to insert them in the soil whilst the sap is still in fresh condition. If the cuttings are left lying about and the sap dries, they will never root.

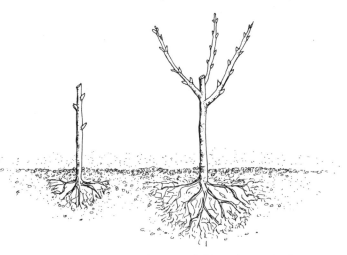

Fig. 3 Gooseberry cutting taking root (left) and new bush showing the 'leg' (right).

Even fresh cuttings will need help to root. Before planting them, insert the severed end in hormone powder and so that it sticks, wipe the end (about 1 in.) with a damp cloth. Before doing so, remove all but the top three or four buds so that the cutting, which should be about 6 in. long, will have a 'leg'. Then insert it into soil. This may be in a garden frame or in the open ground.

The method is to make a V-shaped trench about 9 in. deep. This is filled to about 2 in. below the top with a mixture of sand and peat and made firm. This will encourage the cuttings to root more quickly than if planted in soil. The cuttings are now planted 3 in. apart and 1 in. deep, so that the hormone-treated end is just covered. Make the cuttings firm and water them in, keeping the rooting medium comfortably moist. The cuttings will root quicker if the rows are covered with cloches. If barn cloches (as against tent cloches) are used, a double row of cuttings can be rooted alongside each other, for barn cloches are nearly 2 ft wide.

Rooting one's own cuttings is an inexpensive way of increasing the stock. Cuttings required should be removed before any pruning is done in winter. After removing the shoots, little other pruning may be necessary.

The cuttings will have rooted by the following summer but must always be kept moist. If cloches have been used, they can be removed in early April and used to cover strawberries. It is then advisable to replace some of the soil removed from the trench so that it will cover the peat and sand to a depth of 1 in. This will prevent too rapid evaporation of moisture during summer and will also provide the surface roots with food. Allow the cuttings, now rooted plants, to grow on until October, then lift them with care, so as not to sever the roots, and replant into prepared ground where they will grow on to fruit in two years' time. Until they grow large, strawberries can be grown between the rows. Or the gooseberries may be planted 2 ft apart and alternate plants removed and replanted when about four years old.

When obtaining stock plants from nurseries specialising in soft fruits, obtain two to three year old plants which have not become too acclimatised to the conditions of the nursery. Older plants will resent moving and may take several years to re-establish themselves.

Gooseberries are mostly grown as bushes but the dessert varieties, where size of fruit is all-important, are equally suited to growing as single or double cordons. Where space is limited, they may be grown alongside a path in a row or against interwoven fencing, tied in to strong wires. Plant single cordons 12 in. apart and at an oblique angle for they fruit better when the flow of sap is restricted.

A single cordon is trained by cutting back all lateral shoots to a single bud. This forms the extension shoot which is grown on to an indefinite height but usually no more than 4–5 ft. New shoots formed in summer are pinched back to within 2 in. of the base or main stem in late July or early August, after fruiting has finished.

A double cordon is made by cutting back the leader or extension shoot to two buds about 8–9 in. from soil level. The buds should be on either side of the stem and from here new shoots are formed. These are first trained at an angle of 45°, held in place by two short canes fastened to the wires, then horizontally. When they have grown to about 7–8 in. on either side of the 'leg', they are cut back to an upwards facing bud whilst all other buds are rubbed away. From the two buds, the shoots grow upwards, parallel to each other and about 12–14 in. apart. After the first summer, they are cut back to one bud about 6 in. above the horizontally formed shoots and from this bud the extension shoots are formed. These may be grown on to 4–5 ft high, being treated like single cordons. Red currants may be treated in the same way. Make sure to tie in the cordons to the wires as they continue to grow, otherwise in high winds they may break off and several years' growth will be lost.

Gooseberries may also be grown as standards, like roses, those varieties of pendulous habit being suitable. They crop heavily and at the same time are most attractive when fruiting. Specialist growers occasionally offer standards which should be planted 5–6 ft apart and are suitable to use alongside a path. The plants are whip-grafted onto stems 4–5 ft tall. They will need careful staking and tying for the stems are thin and the heads heavy when in fruit and after rain. They are also easily blown over by strong winds and should not be planted in exposed places. They can be grown in rows between bushes which are planted 6 ft apart which will increase the weight of

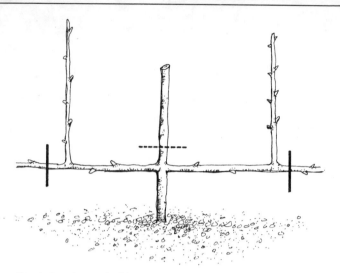

Fig. 4 Forming a double cordon. When formed, shoots have their laterals cut back to a single bud, no other growth being allowed.

fruit from a limited area. The heads are pruned in the same way as for bushes.

Culinary varieties may be gathered at almost any time over several weeks from the time they have made some size and have begun to colour, but dessert kinds should be removed when in peak condition, like pears and peaches. If left too long, they will be over-ripe. The berries should be quite large and well coloured but must not have begun to split their skins. They should be firm but not hard and be sweet and juicy, being at their best with a glass of Madeira, which brings out their subtle flavour, after the evening meal. Those who have not tasted gooseberries in this way have missed much.

Gooseberries are coloured red, white, green or yellow when fully ripe. All are equally delicious though red varieties have a sharper taste as they contain rather more acid. If growing for dessert, only one plant of each variety need be grown but select them to fruit over as long a period as possible. For culinary use, several bushes of one variety should be planted, to give a generous amount of berries at any given time.

Varieties

RED

Bedford Red. Makes a plant of neat upright habit and ripens early, the large round berries turning deep crimson. They are of good flavour.

Dan's Mistake. A chance seedling, it makes a big spreading plant and ripens mid-season, the large oval berries being mostly borne on the older wood.

Lancashire Lad. Raised in 1824, it makes a large spreading bush and is useful in that the early fruit can be picked green for cooking and later, when it has ripened to deep red for dessert. It shows marked resistance to mildew.

London. Produces a large smooth crimson berry, in fact the record holder for weight of any gooseberry and is of delicious flavour but the plant makes little new wood unless grown well, needing plenty of nitrogen.

May Duke. Of neat, upright habit, it is the first to mature and may be picked green at the end of May and when crimson, late June for dessert.

Warrington. One of the last to mature, the bright crimson berries long retaining their firmness so that this variety is excellent for jam making and freezes well.

Whinham's Industry. Makes a large, spreading plant and crops heavily in mid-season, the crimson berries being used for cooking and later as dessert.

YELLOW

Bedford Yellow. A mid-season variety, the large

berries being golden yellow, streaked with crimson and of exceptional flavour.

Broom Girl. One of the first yellows to ripen, it is good for culinary use and dessert, the golden fruits being shaded with green, and sweet and juicy.

Early Sulphur. Ripens by early June; the pale yellow fruit makes delicious eating. It makes a large, spreading bush and bears heavily.

Laxton's Amber. Of compact, upright habit it is a mid-season variety which bears heavy crops of medium-sized berries of golden-amber.

Leveller. Making a large bush, it crops well if given generous supplies of nitrogen and potash. Its huge golden fruits, which mature late, are without equal for dessert.

New Gem. Of upright habit, it is a fine all-round variety, good for kitchen use when not quite ripe whilst later it can be used for dessert.

GREEN

Drill. Making a small compact bush, it is one of the last to mature, the large berries ripening to bottle green and with excellent flavour.

Green Gem. Introduced by Laxton Bros, it is valuable for bottling and freezing when picked early and for dessert when fully mature.

Green Overall. One of the most delicious of all, having a muscat flavour. The huge, bright-green, round berries are covered with grey down.

Gunner. A late mid-season variety, it is one of the best for kitchen use and later for dessert. The large, round, dark-green berries are veined with yellow.

Howard's Lancer. If not susceptible to mildew it would be the best ever raised for it bears heavily in all soils, the fruits being good for bottling, freezing and dessert. The last to mature in August. E. A. Bunyard, the famous nurseryman and writer on good food, said: 'Its fine flavour, enormous crop and sustained vigour are remarkable.'

Thumper. The large berries with their grey-green skin devoid of hairs are one of the best of all for dessert when ripe in July.

WHITE

Careless. The almost oblong berries are of greenish-white when ripe and retain their quality after more than a year in the deep freeze. Good for all purposes, it crops heavily in all soils.

Keepsake. The earliest white, it has all the good qualities of Careless and is good for all purposes. Like Careless it makes size early.

Langley Gage. Of upright habit, it bears large crops of small transparent fruits of quite exceptional flavour.

Whitesmith. One of the earliest and best varieties, it makes a big upright bush, and bears its large downy berries along the whole length of the branches.

Pests and diseases

Aphis. A pest which attacks gooseberries and black-berries, also peaches and plums. The eggs winter on the plants and if they hatch, the aphis will feed on the young shoots and leaves, causing them to curl, turn brown and then fall prematurely, undermining the plant's vigour. The pests will also attack the young fruits. To control, spray the plants in early summer with Abol X or Sybol or with Lindex or liquid derris and repeat at monthly intervals until the fruit forms and continue the treatment after picking until the autumn. Do not use Lindex or any pesticide based on gamma-BHC on blackcurrants. In winter, use a tar-oil preparation on gooseberries and currants.

Gooseberry Mildew. A disease which attacks the foliage, then the fruit, covering it with white mould. Later, it attacks the branches with a brown growth which peels off in winter. It causes stunted growth and the fruit fails to mature. Spraying with lime-sulphur in early spring and again in October will give control but it must not be used on yellow varieties or it will cause bud-drop. Keepsake and Howard's Lancer, however, will readily take lime-sulphur.

For the other yellow varieties, spray regularly with a solution made up of 4 oz soft soap and 1 lb of washing soda dissolved in 4 gallons of water.

Leaf Spot. This is sometimes troublesome in a cold summer or where excessive rain causes waterlogging and the leaves become covered with brown spots. Control by spraying the plants after the fruit is gathered with Bordeaux Mixture. This is made by dissolving 2 oz of hydrated lime and 1 oz of copper sulphate in 2 gallons of water. As it has a detrimental effect on the fruit, it must not be used until after cropping has ended.

Sawfly. The sawfly lays its eggs beneath the leaves along the veins and, in a few days, caterpillars hatch out and begin to devour the foliage. If not checked, the whole plant will be defoliated and will be so deprived of its vigour that little fruit can be expected the next year. To prevent, spray the plants in spring with derris as the leaves unfold and repeat when the fruit has been picked.

5 Strawberries

The first strawberries
The strawberry is most popular of all soft fruits, enhancing the summer scene with its delicious eating. It comes into bearing within eight to nine months of planting and will bear well whilst other fruits take a

year or more to establish. It may also be grown almost anywhere, even where there is no garden in the accepted sense, in tubs and tower pots on a terrace or verandah. Shakespeare appreciated its qualities. In *King Richard III*, Gloucester says:

My Lord of Ely, when I was last in Holborn
I saw good strawberries in your garden there.

By early Stuart times the strawberry had become so profitable a crop to sell to the London gentry that nurserymen in Southgate grew it as a specialised crop as well as picking the fruit from the hedgerows; women carried the berries to Covent Garden market in large circular baskets balanced on the head. Often the fruit made as much as £10 per pound though highwaymen would lie in wait to rob the women of their takings on the journey home. But at such a price, the risk was evidently worth taking.

The strawberries they carried were those which also grew wild about the hedgerows, the native *Fragaria vesca*, a plant that grows well over all north Europe. Oliver de Serre, in the *Théâtre d' Agriculture* (1600) tells of those places, e.g. woodlands and hedgerows, where wild strawberries are found. Another less common wild variety is *F. elatior*, the Hautbois, beloved by French connoisseurs of good eating in the nineteenth century and so called because it holds its fruit high above the foliage. About this time the Alpine or Runnerless strawberry was introduced, the earliest variety being Baron Sole-macher, the fruits being small but the flavour outstanding.

It was the introduction of *F. virginiana* from America, probably by John Tradescant in the seven-teenth century that gave us a strawberry of any real size. Shortly afterwards, a French naval officer brought back plants of another and much larger fruiting variety which he found in Chile where it grew wild and right up the western seaboard of America to Alaska. It was named *F. chiloensis* and was found to be sterile until fertilised with pollen of *F. virginiana*. Perhaps the old Hautbois was also used in the hybridising for soon after there appeared in France the variety Ananas (pineapple flavoured), which had much of the flavour of the Hautbois.

Modern varieties

Early in the nineteenth century, Michael Keens was working at Isleworth with several of the early forms. In 1820 he introduced to commerce a plant which he named Keens' Seedling. In both size and flavour it was far in advance of anything previously seen in England and it set the standard for the future. In addition, the plants were of robust constitution and cropped heavily. But it was Keens' Imperial which preceded it that was to achieve greater fame, not for the fruit but it was from a seed of this variety that John Wilmot, in 1822, introduced the famous Black Prince, a strawberry which was entirely devoid of any acidity and treacle-sweet. It was this variety that Thomas Laxton used for pollinating the popular French introduction, Viscomtesse de Thurty, which in 1888, gave us King of the Earlies. This in turn was the most popular variety of its day and was the pollinator (with Noble) for Laxton's magnificent Royal Sovereign, named in honour of Queen Victoria and which appeared in 1892. It has never been surpassed either for flavour or appearance and is still the most popular strawberry grown.

Surprisingly, out of all the Common Market countries, strawberry consumption in Britain is by far the lowest, only 3 lb per head in a year compared with more than double that figure elsewhere in Europe. Owing to the high cost of picking the fruit, the strawberry acreage in Britain is now only about

10,000 acres—half of what it was twenty years ago. Soon, we may be forced to grow our own if we want to enjoy fresh strawberries.

It is usual to have our strawberries with cream and sugar, though a sprinkling of Beaujolais on the fruit just before serving will bring out the flavour far better and not more so than with the Cambridge varieties Rival and Aristocrat.

General considerations

Much more susceptible to frost damage than the gooseberry, the strawberry also needs to be picked the moment it is ripe. It cannot be allowed to remain on the plants a moment longer, unlike the gooseberry which can hang for a week or more, until there is time for its picking, without deterioration. In addition to frost, strawberries may also be spoilt by heavy rain when reaching maturity and in pre-war days this often meant the ruination of the entire crop. Today, it is possible to plant for succession so that the fruit may be enjoyed from mid-April to early May (under cloches) then from early June until August and until Christmas from the autumn-fruiting varieties, again under cloches.

Some varieties such as Royal Sovereign and Cambridge Early Pine, which have a glossy skin surface, are more able to shed moisture than others and should be planted in areas of high rainfall. These two varieties also show marked resistance to mildew, which in some years attacks plants growing in a warm, moist climate. But under average conditions in the Midlands and for a long succession of fruit this would be a suitable selection:

Cambridge Premier (cloches)   April–May
Cambridge Early Pine   Early June
Cambridge Premier   Mid-June

Cambridge Regent (more frost resistant)	Mid-June
Royal Sovereign	Late June
Fenland Wonder	Late June
Talisman	Early June
Cambridge Late Pine	Mid-July
Cambridge Rearguard	Late July–August
Hampshire Maid	August
Gento	September–October
La Sans Rivale	September–November
La Sans Rivale (under cloches)	November–December

If space is strictly limited, then choose from Cambridge Early Pine, Royal Sovereign, Talisman and Gento.

Possibly the best for early forcing under cloches or in a cold frame is Cambridge Premier. This variety makes only a small amount of leaf and is untroubled by mildew. But if the plants are to be covered, they must be planted before the end of September so that they become well rooted whilst the soil is still warm. Those which fruit early when unprotected, e.g. Cambridge Premier and Early Pine, will also require planting at this time for the same reason and will then crop only eight months after planting. Those which mature later, e.g. Cambridge Late Pine and Hampshire Maid, may be planted in October and November and any of the autumn-fruiting varieties during the latter month, but all strawberries expected to fruit the following year must be planted before 1 December when frosts can be expected. The plants must be well established before the frosts can do their worst. Spring planted strawberries should not be allowed to fruit until the following year and during the first summer should have the blossom removed, cutting away the flower trusses with scissors.

One is often asked how much fruit can be expected from the plants. The home grower can expect to obtain from four rows 25 ft long, each of which will take twenty-five plants, about 100 lb of fruit if the conditions are good and the plants are two years old. This would make a suitable plantation in the average garden, planting for succession and confining each row to a single variety. This will give worthwhile pickings whenever it is decided the fruit is sufficiently ripe. Even without cloches, there would be fruit from early June until late September and only a limited amount spoilt by adverse weather. The plants will reach a peak of production in their second year with rather less fruit in their third and fourth year when the quality will also not be quite so good.

Preparing the ground

Strawberries will crop well in a variety of soils, but unless well supplied with humus will rarely do well in a limestone soil, Cambridge Brilliant and Vigour being the exceptions. Strawberries enjoy best a deep loam and will do well in light land if plenty of humus is incorporated to retain summer moisture. Heavy land must be well drained otherwise the plants will decay at the crowns in winter. A heavy, compact soil will never grow good strawberries; they like a deeply worked soil which is in a friable condition otherwise they will have difficulty in sending out their fibrous roots in search of food and moisture. But of all fruits, they will flourish in a soil of quite high acidity. Probably the highest yields at present are obtained from plantations in parts of Ireland, in Co. Wexford and Sligo, where the soil is of a peaty nature and shows a pH of 5.0–5.5 with neutral at 7. Here, as much as thirty tons of farmyard manure is ploughed into every acre and in some years the yield reaches ten thousand pounds of top quality fruit, double the average for other parts of the British Isles.

In early autumn, dig the ground at least a spit (spade) deep, incorporating whatever humus is obtainable. Clearings from ditches, shoddy, decayed manure, and straw which has been composted with an activator and some poultry manure, are all suitable. They will provide the necessary plant food and will lighten the soil. Strawberries also love peat and it is rarely possible to give them too much. Peat opens up the soil and provides the plants with the slightly acid conditions needed for high yields. Where peat can be used in bulk, then add only small amounts of other humus-forming materials.

Above all, strawberries need clean land. They will not do well in weed infested soil and will soon give up the struggle for existence. Nor is it possible to clean the ground after planting; the plants soon begin to send out runners making weeding, unless by hand, almost impossible and to hoe too near the plants will disturb the roots causing them to die back. It is also not advisable to plant in newly dug turf-land for here, wire worm so often proves troublesome, destroying the roots. If the ground has previously been turf, treat it for wire worm accordingly.

Prepare the ground well in advance and allow it several weeks to settle before planting. If peat has not been used in quantity, rake a generous amount into the top soil so that the plants' roots will be in contact with it at planting time. Also rake in 2 oz per square yard of sulphate of potash just before planting. Apart from this, other artificial fertilisers should not be used.

Planting

There are many ways of planting strawberries. The Fenland growers plant in rows on the flat and where the land is well drained this can be advised. The rows are made 2 ft apart with 15 in. between the plants in the rows. They can be covered with barn cloches or a continuous tunnel cloche of plastic material. If the ground is of a heavy nature and not too well drained, ridges are made the same distance apart (as for potatoes, earthing up the soil to a height of 6–8 in.) and the plants are set out along the top.

Experiments carried out at the Botley Experimental Station in Hampshire have shown that where planting was done on ridges, far higher yields were obtained, although in this case the land was well drained. Also, on deeply cultivated and well-manured land, not only was the yield double that from plants growing in badly cultivated land but growth was so strong that the plants continued to bear heavily for four and five years whilst those from poor land cropped for only two years.

In badly drained, low-lying land, the plants may suffer from Red Core disease, which causes the roots to turn red and die back and the plants will decay at the crown. Planting on ridges, if the soil is heavy or not well drained, will help to avoid this.

If there is a risk of frost, again plant on raised beds. Those varieties that are resistant are: Cambridge Early Pine; Regent; Favourite; and Vigour. Also resistant is Spangsbjerg, a Danish introduction.

Raised beds are made 5 ft across and of any length. This will enable the plants to be picked over from both sides of the bed without treading the soil. If planted 15 in. apart in rows made 18 in. apart, four rows can be planted to each bed. With this closer planting, the foliage will give some protection against frost and cold winds. In their second year, the plants should be allowed to keep their runners which will bear fruit the following year. The plants should be removed after four years and fresh beds made up. Where growing in this way, the soil must be clean and

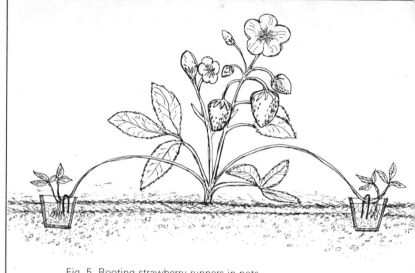

Fig. 5 Rooting strawberry runners in pots.

well fortified with organic manures.

Plants growing in rows should have their runners removed when they have formed roots. Runners are formed on string-like attachments which spread out from the parent plant to about 10 in., forming new plants which take root. From these plants, others may develop but it is the first one that is the strongest and should be used to make a new plantation. Others which may grow on from the newly rooted plant are discarded. The parent plants, however, must not be allowed to form too many runners, otherwise they will rob the parent of much of its vigour and crop yields will diminish.

Plants can also be rooted directly in small pots filled with a friable compost. The pots are inserted in the ground as shown in the diagram.

Always obtain 'maiden' plants which are from certified stock, the grower having received a certificate from the Ministry of Agriculture guaranteeing his plants as free from virus disease and Red Core. There are specialist growers catering for the plant trade who have a wide selection of varieties and who sell only certified plants. They may cost a little more but are worth it. To begin with, clean, healthy plants go a long way towards success. 'Maidens' are taken from year-old plants and are the most vigorous.

Plant when the soil is in a friable condition, which it usually is in autumn, and until ready to begin planting, keep the strawberries away from drying winds. Use a blunt-nosed trowel and make the hole large enough for the roots to be spread out before covering them in. Strawberries planted with the roots bunched together will never make good fruiting plants. Do not plant too deeply—the crown of the plant should be at soil level—and tread in firmly. Like most soft fruits, strawberries are surface rooting. A garden line should be used to space the rows correctly and to set the plants straight in the rows. This will facilitate later cultivations such as hoeing between the rows, which can then be done without going too close to the plants. Make the rows north to south to ensure the even ripening of the fruit.

Care after planting

The plants will require no further attention until early March when it is advisable to look over the rows and to tread in any plants made loose by frost. Then hoe between the rows, but not too near the plants, to break up the surface soil which may have become consolidated by winter rains. At this time and only if the ground is in good 'heart', give the plants a 1 oz per square yard dressing of sulphate of ammonia to stimulate them into growth. For plants in beds, give double this amount but only where the soil is well supplied with humus. It is best given on a showery day so that it will be washed into the soil.

By early April the plants will have begun to form new leaves, but if the weather is dry, water them occasionally to keep them growing. By the month end, the plants will have begun to form their flowering trusses and they should be mulched between the rows to conserve moisture and to feed the surface roots. Peat, mixed with a little decayed manure, is ideal. It should be worked right up to the crown of the plants for this is where the mulch is most needed.

Protecting the plants

Early May, whilst the plants are forming their first blossom, is a time of apprehension for the fruit grower. The blossom may be seriously damaged by one night's hard frost though the frost resistant varieties will often come through all but the hardest frosts unharmed. Where cloches are available, they should be placed over the plants at the end of March, though not before as the plants must not be forced into bloom before they have made some leaf. Covering too soon is often the cause of failure where growing under cloches.

It is usual to plant a row of an early variety for cloching and to cover the entire row of not less than a dozen plants so as to give worthwhile pickings. Where barn cloches are used, they will cover a double row spaced about 16 in. apart, but be sure to have the plants in the ground by the first days of September to give them at least six months to form a strong rooting system before they are covered. The best for cloching, in order of ripening, are:

1. Cambridge Brilliant   4. Cambridge Rival
2. Cambridge Premier   5. Cambridge Vigour
3. Cambridge Profusion   6. Royal Sovereign

Make sure that the cloches are clean before they are in position. Before covering the plants, give them a soaking and then dust the foliage with flowers of sulphur to guard against mildew although these are resistant varieties.

By early April the earliest varieties will be showing bloom and to help with pollination remove the covering on mild days. At this time, syringe the plants and if the soil is dry, give a soaking before replacing the cloches late afternoon, before it turns colder. The ventilation given will do much to prevent mildew but do not uncover the plants if cold winds are blowing for in a few hours the winds will undo all the good the cloches have done. If one has to be away from the garden during daytime, then on calm days, space out the cloches 1 in. apart to allow insects to pollinate the flowers and close them up when returning each evening. Then if cold winds arise during the day, the plants will take little harm. But if in doubt, play safe and keep the plants fully covered.

The first fruit will be ready about mid-May when the weather will become warmer, which is the signal that the fruit will be in good demand. There will be no need to straw the plants to prevent soil splashing onto the fruit for they will be protected by the cloches. Keep them in place during showery weather as the fruit ripens, for strawberries are better picked dry. The cloches can be removed for several hours after each flush of ripe fruit has been picked.

The earliest varieties of uncovered plants will begin to form their fruit towards the end of May. It is surprising how quickly the tiny green berries turn white and begin to swell, taking only ten to twelve days to form the large red fruits which should then be gathered at once. It should be said that speed in ripening will depend entirely upon the weather though situation will also play a part; in Somerset, where it is grown commercially around Cheddar, the fruit reaches maturity at least ten days before that grown in Kent or the Midlands, and often fully a fortnight before that grown in the north.

As soon as the fruits have formed, it is advisable to

give protection from splashing. Strips of black polythene, 10–12 in. wide, may be placed between the rows, close to the plants and held in place by stones. This will prevent soil splashing on to the fruit and being black, will help the soil to retain the warmth of the sun and the plants will reach maturity more quickly.

Where plants are growing in beds and the rooted runners grown on to fruiting, it is difficult to give any protection apart from a peat mulch. Where in rows, the soil may be covered with straw right up to the plants and to a depth of about 6 in. The straw is removed and used for composting when the crop has finished. The use of straw to keep the fruit clean gave the plant its popular name during the Middle Ages. There are, however, several adverse points in using straw—it tends to harbour pests and retains frost.

Where obtainable, special mats, made of whale-hide, plastic or bamboo, should be used by the private gardener. The mats are made 12 in. square and at the centre is a hole of 3 in. diameter through which the foliage of the plant is put, the mats then lie flat on the ground. Some varieties tend to bear their fruit in short trusses at ground level and need more attention to the protection of the fruit than, say, Cambridge Regent and Rival which hold their fruit well above the foliage, like the old Hautbois.

As an inexpensive alternative, single sheets of news-paper, folded over and with a hole of 3 in. diameter cut out at the centre, may be used in place of mats or straw. The paper 'mats' are held in place with stones and will last for several weeks, until the fruit has ripened. The paper will then act as a useful mulch. Broken crocks, glass or stones may also be used to prevent soil splashing.

If a period of dry weather is experienced, and this often happens in late May and through June, water the plants regularly until the fruit begins to ripen. If dry at the roots, and the straw or mats will encourage this, the fruits will be small and seedy. This is why it is so important to provide the soil before planting with as much humus as possible to reduce the need for watering. Plants which are dry will often be troubled by red spider, especially where in a light soil. Plants growing in such a soil will need more moisture than those in a heavy soil. These varieties, in order of ripening, fruit well on light land:

1. Cambridge Early Pine
2. Cambridge Premier
3. Cambridge Regent
4. Cambridge Favourite
5. Cambridge Vigour
6. Talisman

After the first fruit has ripened, the plants should be given a watering with dilute liquid manure (as for gooseberries) once a week, preferably during showery weather so that it is quickly washed down to the roots and absorbed by them. It will prolong the crop, improve the quality of fruit and help to build up a sturdy plant for next year's fruiting.

Remember that when once the fruits have swollen and turned white, they will ripen quickly and the plants should be picked over every day. Handle the fruits with care and place them in a basket or on a tray, then take them indoors as quickly as possible. Remove the green tops and, to improve their eating, place in the refrigerator for an hour before using.

For deep freezing, strawberries require rather more attention than other fruits. The method is to place the berries in the freezer on trays with the berries not quite touching each other. Give them two hours to become frozen solid, then remove them, put them in plastic bags and replace them in the freezer. If this method is followed they will retain their shape and eating qualities for a year and will not go mushy when removed from the freezer and used.

After the plants have finished fruiting, remove the

runners, if growing in rows, and use those nearest the parent plants to make another bed, as described. Remove any straw and, if possible, give the plants another mulch which will gradually be worked into the soil during winter. The foliage will die down gradually and by the year end there will be little sign of the plants being there. They will begin to grow again with the first warm days of spring.

Growing in tubs and barrels
Excellent crops of strawberries may be obtained from tubs made from barrels sawn in half; or full barrels may be used with the lid removed. Drainage holes must be drilled in the bottom and at intervals of 12–15 in. around the side so that many more plants may be inserted, in addition to those planted at the top.

Tubs and barrels of any size may be used but the larger the better, as more plants can be accommodated. A tub of 24 in. diameter is a suitable size and can be placed on the sunny side of a courtyard or on a terrace or verandah and, if there is room, several tubs can be used. They are obtainable from the Barrel House, St Agnes, Cornwall, a company which specialises in them and will drill them ready for use; or they may be obtained from cider or vinegar makers. They should be treated with a wood preservative before filling and the iron bands should be treated with a rust-proof preparation before painting them black. Being of oak, they will then last for many years in good condition.

Put the tubs in place before they are filled as they will be too heavy to move afterwards. First, line them at the bottom with crocks or broken brick to prevent soil blocking the drainage holes, and over this place a layer of turf, grass side downwards. Then fill up the tubs with prepared compost. This should consist of

Fig. 6 Strawberries growing in a barrel.

good quality loam to which has been incorporated some peat and used hops or decayed manure. Work them well into the soil so that those plants set in the side holes will have their roots in rich fare. Allow the compost several days to settle. The tubs should be filled to about 1 in. below the top to allow for watering without the soil splashing over the side.

As a guide, to every barrow load of soil add a bucketful of peat, which must be quite moist, plus half a bucket of used hops or decayed manure. A handful of bone meal worked well in will also be of value. October is a suitable month to prepare the tubs and if planting is also done at this time, one may expect to have fruit the first season. Plant only 6 in. apart and to insert the roots into the compost through the holes use the end of a strong spoon and press them in.

From early April, the plants will require regular watering, giving sufficient to reach down to those plants growing from the holes for they will receive little natural moisture. They will begin to fruit early in June and an occasional watering with dilute liquid manure will enhance the quality of fruit and build up sturdy plants for the following year.

Fig. 7 Strawberries growing in a tub.

Where space is still further restricted, strawberries may be grown in earthenware pots (one plant to a 6 in. pot) which may be placed in rows where the sun can reach them. Use a similar compost as for tubs but they will require frequent watering in summer as the compost will dry out quickly.

Tower pots will take up even less room and may be stacked to any height. They are made from heavy-duty white plastic and are almost unbreakable. The pots lock together to form a pillar and, to prevent their being blown over, may be fastened to a wall by means of a strap looped around the pillar. The height of twelve pots is 5 ft 6 in. and the pots have a diameter of 5–8 in. Each pot has two 'balconies' opposite each other and here the plants are set. Use a similar compost as for tubs. Manufactured by Ken Muir of Wheeley Heath, Clacton-on-Sea, a set of six with a drip tray at the base, costs about £5 and may be used year after year with strawberries as a permanent crop.

Varieties

EARLY

Cambridge Brilliant. This crops well in limestone

soils and as it makes a compact plant with not too much foliage, it is suitable for cloches, bearing its bright scarlet wedge-shaped fruit in short trusses.

Cambridge Early Pine. Though highly resistant to mildew, it is rather too leafy for cloches. It is excellent for wet districts for the rain quickly runs off its glossy fruits, which are round and bright scarlet. The blossom is frost resistant.

Cambridge Premier. Resistant to mildew though not as frost resistant as Early Pine and Regent but making a more compact plant; it does well under cloches. The bright-orange wedge-shaped fruit is firm and freezes well. It crops well in heavy soils.

Cambridge Regent. One of the first to ripen, it is the hardiest, being highly resistant to frost. Whilst it does well in all soils, it prefers a heavy loam. It should always be grown in the open as under cloches it is often troubled by mildew. The scarlet wedge-shaped fruit is firm and freezes well.

Hative de Caen. A hardy, frost-resistant variety from northern France, making a big leafy plant and maturing early; the large, round, deep-pink fruits have a rich fragrance.

Reine des Précoces. With its dark foliage and bright red fruits, it is handsome when cropping and being resistant to mildew does well under cloches. Otherwise, grow it in the warmer west.

SECOND EARLY

Cambridge Favourite. Troubled neither by frost nor drought, neither by mildew nor botrytis, it is one of the 'fool-proof' varieties which crops well in all soils and in most years. The commercial growers' favourite. It is the best for a light, humus-lacking soil and bears its large red fruits over a longer period than any other variety.

Cambridge Rival. Like Regent, it does best in a heavy

loam and in the west is more popular than any other strawberry, the bright-red conical fruits hanging above the foliage in large trusses. Though making a large plant it does well under cloches as it is not troubled by mildew. Its fruit was chosen by 120 members of the RHS Fruit Committee as having the best flavour of all.

Cambridge Vigour. Resistant to Red Core and mildew, it is of strong constitution and crops well even in a limestone soil. The glossy deep-crimson fruit is of medium size and freezes well.

Grandee. Produces the largest berries, weighing 2–3 oz when grown well, and loses nothing of its excellent flavour in achieving this size.

MID-SEASON

Cambridge Sentry. An excellent mid-season variety for a heavy soil and for wet districts as the glossy crimson wedge-shaped fruit is untroubled by rain and being held well above the ground is protected from soil splashing by its foliage.

Huxley's Giant. Introduced in 1912, it has since remained popular with market growers for it is of strong constitution and is troubled neither by frost nor disease. The large, irregular crimson fruits ripen slowly but the yield is good and the plants crop well for three or four years.

Red Gauntlet. The most reliable mid-season variety, it is a compact grower with neat foliage and holds its fruiting trusses well above the ground. The large scarlet fruits have good flavour and it crops well in all soils.

Royal Sovereign. Never equalled either for flavour or appearance, the brilliant scarlet wedge-shaped fruits retaining their shape and quality in the deep freezer. The virus-free Malling 48 strain is the best and this variety is best grown in the drier east of Britain. It requires a light soil enriched with plenty of manure.

LATE

Cambridge Late Pine. One of the last to ripen, its large crimson fruits are the sweetest of all strawberries and it crops well in all soils.

Domanil. Very late, it crops heavily in all soils, the large conical berries being bright orange with brisk orange flesh.

Fenland Wonder. The original strain was found growing on a church wall at Emneth in East Anglia. It has so many good qualities that it has been widely planted in that part of the country. The large crimson fruits remain firm after freezing and it crops well over a long period.

Hampshire Maid. The last to ripen, it bridges the gap until the autumn-fruiting varieties are ready. In a well fortified soil it crops heavily over a long period, its large conical fruits ripening to deep crimson and being sweet and juicy.

Talisman. Raised in Scotland, it shows resistance to Red Core and crops heavily over a long season. The scarlet wedge-shaped fruit retains its shape after freezing and is of good flavour.

Autumn-fruiting strawberries

These have been grown on the Continent for many years but for some reason are only now becoming popular in England. They are hardy—the blossom of La Sans Rivale withstanding many degrees of frost—but away from the south-west require cloche covering to ripen the fruit. Then in a good autumn and early winter, they will continue to ripen until Christmas.

They are planted out in well-nourished and well-drained soil towards the end of March and where covering with barn cloches, it is usual to plant in a double row 18 in. apart and 15 in. in the rows. The plants are allowed to form runners which will bloom and fruit in late autumn, with the parents. La Sans Rivale and Gento are especially prolific with their runners.

The plants require the same rich soil and cultivations as for the summer varieties. In particular, they must have a soil containing plenty of humus to retain moisture, otherwise the fruit will be small. During summer, water copiously. Until the end of June, remove the blossom so that the energies of the plants will be directed into autumn fruiting.

About mid-September, cover the plants, removing the cloches on all warm, sunny days. Covering them when the autumn days are wet and foggy will protect the plants from mildew. As an extra precaution, it is also advisable to dust them with sulphur and if the soil is dry, water only the roots at this time.

The plants will begin to fruit late in September and by the end of winter, when the cloches are removed, will have died down completely. In spring, give the beds a generous mulch of decayed manure and peat and begin watering if the weather is dry. Last year's runners will now begin to produce runners and by the end of summer the ground will be a mass of plants all ready to fruit. Water frequently with liquid manure so that the plants do not impoverish the soil; this also helps the fruits to grow large and juicy.

Varieties

Gento. From September until mid-November it bears heavily and will do so in its second year. It forms numerous runners to continue the cropping year after year. The conical light-red fruits are large and firm and of excellent flavour.

La Sans Rivale. An amazing variety—I have had it fruiting outdoors in North Staffordshire in frost and snow on Christmas Day. Under cloches the fruit ripens better, continuing from early autumn until early January, the conical berries colouring pale red and have good flavour.

Rabunda. A new Dutch variety, remarkable for the enormous crop it gives in early autumn, the large bright-red berries being firm and juicy.

Alpine strawberries

The alpines, *Fragaria vesca*, enjoy some shade and may be planted between gooseberries or in those less sunny parts of the garden where other strawberries would not ripen so well. The alpines (and they originated in the alpine regions of France and Switzerland) may be grown in pots and window boxes as they make compact plants 6–8 in. high and do not form runners. They are raised from seed and increased by division of the crowns in spring. The fruit is small but makes a delicious conserve and as a sweet course, sprinkled with red wine or claret, it is indeed 'food for the gods'. Or mix a few with the less fragrant summer kinds to give them additional interest.

Grow them as an edging to a path or border or in small beds to themselves. They require a soil containing plenty of humus to retain summer moisture, otherwise the berries will be small and hard. Keep the plants well watered and give them an application of liquid manure every ten days. They will also appreciate a liberal mulch of peat and decayed manure in June.

The main flush of fruit comes late in July and continues until the end of August but the fruit will continue to ripen until the end of October.

The alpines grow true from seed. In April sow them in boxes or pans containing John Innes compost. Either place the boxes in a frame and keep the lights on until after germination, or place the boxes in a sunny corner and cover with a sheet of glass. Keep the compost comfortably moist and as soon as the seedlings are large enough to handle, transplant them to a bed of prepared soil or to a frame where they can be kept growing until July when the plants are set out in their fruiting quarters. Plant 18 in. apart as they may be left for three years, by which time they will have grown large and bushy, each plant yielding about 2 lb of fruit each season. After three years, lift and divide, re-planting into fresh ground which has been well prepared. Do not plant too deeply—the crowns should be at soil level—but spread out the masses of fibrous roots. Tread them in firmly.

Varieties

Baron Solemacher. One of the best alpines, the bushy plants growing 12 in. tall. It begins to fruit early July continuing until the frosts. If the soil is well nourished, the fruits grow larger than those of other alpines. They are long and rich crimson.

Belle de Meaux. The last of the alpines to fruit, at its best when the others have passed their peak. The crimson wedge-shaped fruits are of excellent flavour but, as with all alpines, should not be gathered until fully ripe or they will be hard.

Delight. The fruits mature creamy-white and are sweet and juicy. They should be served with the red alpines. They also make an excellent conserve.

Pests and diseases

The strawberry suffers from more diseases than any other soft fruit but the problems are usually caused through poor culture, the use of too many artificial manures in a soil lacking humus, and adverse soil conditions due to poor drainage. Hereditary factors caused the destruction of the fine variety, Climax, and in an effort to provide growers with virus-free stock, the Nuclear Stock Association was formed after the war. Under this organisation, guaranteed virus-free selected clones or strains are built up and awarded a special certificate.

To obtain this stock, selected plants are kept for ten days in a clean room where a temperature of 100 °F is maintained. This is known to kill any virus. The plants are then grown on and propagated under controlled conditions and the resulting stock is awarded a special certificate.

Aphis. The most troublesome of strawberry pests. These tiny insects feed on the sap of the plants causing loss of vitality and allowing virus diseases to enter at the points of puncture. To control, spray with Lindex from the time of the unfolding of the new leaves until the first fruits begin to set, or dust with derris.

Blossom Weevil. These pests lay their eggs in the blossom as soons as it opens, the tiny grubs feeding on the centre of the bloom causing it to be unfruitful. The pest is troublesome only when the flowers appear but can do a great deal of damage if not controlled. As routine, dust with derris at this time.

Botrytis and *Mildew.* Both are forms of mildew, botrytis appearing on the fruits as a powdery dust whilst what we call mildew attacks the foliage. This mildew is kept under control by dusting the plants with flowers of sulphur from early April, especially those growing under cloches. Botrytis of the fruit is controlled by using Orthocide dust containing 10 per cent Captan. This has given new hope to those growers who relied on the heavy cropping of varieties subject to this disease. But in areas of heavy rainfall it is advisable to grow those varieties which show marked resistance, e.g. Cambridge Early Pine and Late Pine, Premier and Talisman.

To prevent an attack, dust the plants every ten days from the formation of the first green fruits. When the fruit ripens, dust between flushes, i.e. when the fruit has been gathered and whilst waiting for the next flush to ripen. Dusting with Orthocide will save up to 50 per cent of the crop in damp, humid weather such as is often experienced in July. Orthocide must not be used where the fruit is to be canned.

June Yellows. This is the term given to an hereditary disease which may take more than twenty years to show itself on any variety when all plantations become affected at exactly the same time. The first indication is the yellowing at the edges of the young centre leaves when the foliage, devoid of chlorophyll, dies back. Runners from the same plant are also affected. The trouble may present itself on plants growing in cold, badly drained soils for which reason one Scottish grower known to me will plant at no other time than early April. He does not allow the plants to fruit that season, but removes the flower trusses as they form so as to build up a strong plant. This may also be good advice for those growing in heavy soil in cold, exposed gardens.

If the trouble shows itself, pull up any infected plants at once and destroy them for, on occasions, it may be due to adverse growing factors.

Red Core. This disease is caused by the fungus *Phytophera fragaria* which attacks the roots at the crown or centre of the plant, causing the roots to turn red at the centre and the plants to die back. Badly drained soil is usually the cause and for this reason the Botley Experimental Station suggests growing on ridges. Once the soil becomes infected, the ground will need a long rest period to recover.

In a heavy, badly drained soil plant resistant varieties, e.g. Cambridge Regent, Rival and Vigour, also Talisman and Fenland Wonder.

Tarsonemid Mite. Occasionally a troublesome pest for the mites cluster in the centre of the plant, when growth recommences in spring, where they lay their eggs and are difficult to eradicate. As a precaution, commercial growers immerse all plants for 20 minutes in water heated to 110 °F (43 °C), taking care to watch the thermometer closely for at 113 °F (45 °C) the plants will perish. For the amateur, dusting the plants with flowers of sulphur (as for botrytis) from the time the leaves begin to unfold will usually give control; or spray with a 2 per cent lime-sulphur solution.

6 Raspberries

Basic requirements

Never as widely planted as strawberries though for the amount of ground they occupy, they bear more fruit than any other soft-fruit crop. The canes grow upwards with several canes from each root (or 'stools' as they are called in the gardener's jargon). Ten or twelve canes from an established plant, and occupying no more than a square foot of ground, will bear about 2–3 lb of fruit over a period of three to four weeks.

Rubus idaeus, the botanical name for the raspberry, grows wild over most of Britain and northern Europe but is especially prominent in northern Britain, from Yorkshire to Perthshire in Scotland, for like gooseberries, the raspberry enjoys a cool climate. In the wild it grows in open woodlands up to 600 ft above sea level, usually in a light, damp, peaty soil overlying a gravel bed, which provides it with just the right conditions, i.e. a well-drained soil containing plenty of humus to provide good drainage in winter and plenty of moisture in summer. This is of prime importance for if lacking moisture at the roots, the plants will produce little cane growth. The raspberry is a deciduous shrub which forms a biennial stem or cane—that is to say, it bears fruit on the stems of the previous year's growth.

Until the late Mr Norman Grubb took up the scientific culture of raspberries in the post-war years there were only two recognised commercial varieties, Lloyd George, introduced in 1920 and which is early fruiting, and Norfolk Giant which fruits later. Both are now still widely grown but in addition, the Malling varieties provide fruit over a much longer season. This gave the raspberry a new and much needed popularity for if the early blossom of Lloyd George is killed by frost, there are now others to follow. The season now extends over ten to twelve weeks, finishing with the autumn-fruiting varieties which should be grown south of the Trent so that they will ripen their fruit in an average year.

Several of the Malling varieties may be grown where frost is troublesome. Malling Jewel especially, opens its blossom later than other earlies though it is one of the first to ripen its fruit. Follow with Malling Enterprise, Newburgh and Norfolk Giant, and with the new September to fruit in autumn. These varieties will rarely be troubled by frost.

Possibly, the fact that it is necessary to pick raspberries in exactly the right condition, has told against their popularity for unripe fruit is acid and tasteless whilst over-ripe will be mushy when gathered. As soon as the berries begin to turn red they should be looked over daily, though picking should be done only when the fruit is dry. The berries do, however, possess a flavour and an aroma all their own and come when one may be tiring from a surfeit of strawberries.

Preparation of the soil

Raspberries are tolerant of various soil types and do well in a stiff clay loam over a gravelly sub-soil through which winter moisture can drain. But they best enjoy a light loam well supplied with humus to retain summer moisture. Clean ground is just as essential as it is for strawberries, for raspberries are even more permanent if given good culture—a row of stools (roots) will last for many years, each year sending up a supply of new canes to fruit the following year. Raspberries hate drought, hence natural plantations are always to be found in moist open woodlands where the roots are shaded from the direct rays of the summer sunshine. If growing in a dry soil the fruit will be small and seedy and the plants will produce small, weakly canes.

As the ground is cleared, work in all the humus possible. Decayed farmyard manure, garden compost, bracken or straw rotted down with an activator, clearings from ditches, or used hops are all suitable. In addition, work in some peat which raspberries enjoy. Remove all perennial weeds and work the ground as deeply as possible. Raspberries also require potash, which may be given at planting time in the form of wood or bonfire ash, or give 2 oz of sulphate of potash per yard of each row.

Late autumn is the most suitable time to prepare the ground and for planting the canes so that they will be established before the frosts. Always obtain virus-free stock plants from a specialist grower. Scottish canes are the best and there are growers who specialise in the production of stool beds which are constantly moved into fresh ground, the canes being used to form fruiting plantations which are not disturbed, no canes being removed for the plant trade where growing for the fruit. The canes should be large and sturdy to produce satisfactory plants.

Planting

For most varieties, plant the canes 15 in. apart, or 18–20 in. for the vigorous Malling Promise and Malling Admiral, and do not plant too deep. As the canes are long, it is often the practice to plant them well down, but this is the most frequent cause of failure. Plant so that the roots are only just covered, with the base of the cane just below the surface. Plant in straight lines for the rows will need staking and in the north to south direction so that the sun will reach all parts of the canes and ensure quicker maturity. At least ten or twelve canes should be planted in each row to give worthwhile pickings and make the rows 4–5 ft apart, 5–6 ft for the more vigorous kinds. Three or four varieties maturing at different times will give a long succession of fruit.

Always plant when the soil is friable and tread the plants in well. The best way is to dig a trench 4–5 in. deep then to hold each cane upright and at the same time cover the roots with soil using the feet (suitably

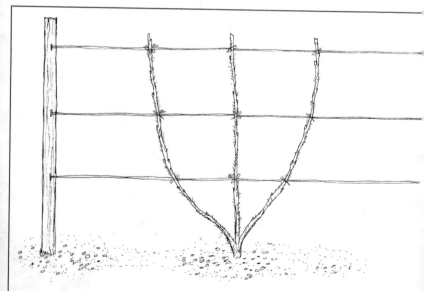

Fig. 8 Supporting raspberries in rows.

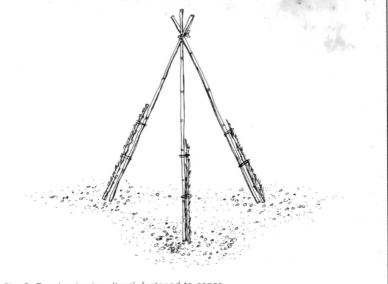

Fig. 9 Raspberries in a 'tent', fastened to canes.

protected by a strong boot or Wellington), the hands being used for planting the canes. Planting can take place from early autumn until the end of March when the canes are cut back to 3–4 in. of their base. This will encourage new cane growth from the 'eyes', which are just below soil level. The new canes will begin to grow at once and by the end of summer will have reached 5–6 ft. They are tied to strong wires which are held in place by stout 6 ft stakes at the ends of the rows, the wire being fixed to the stakes at intervals of about 15 in. As the canes reach the wires, they are tied in so that they will not break in strong winds. The canes will fruit the following year but should they break, a year's fruit will be lost. In spring, it is usual to remove the tips of the canes so as to concentrate on the formation of fruit rather than on further cane growth.

Another method is to plant in circles of 2 ft diameter and to stake the canes in tent fashion. Eight canes should be planted at the circumference and a bamboo cane placed with each after cutting back the canes in March.

Yet another method is to fix 1½ in. square stakes into the ground at intervals of 2 ft and so that the stake is 6 ft above soil level. On either side of each stake plant a raspberry cane and tie the new canes to the stake as they grow. Some growers plant in rows and 'box in' the canes, providing two stakes 12 in. apart at the ends of the rows and taking the wires all the way round the stakes and canes. No tying-in is done but this method is suitable only where strong winds are not experienced.

It is always important to have the support stakes plenty long enough, at least 6 ft above ground with wire at the top, for the canes are brittle and will break at the tops in windy weather if not tied in.

Care after planting

As the new canes begin to grow, they should be given a mulch along the rows. This can be of decayed manure or composted straw, or lawn mowings augmented by some peat. The mulch will help to retain moisture in the soil and suppress weeds. Cultivations should not be taken too near the plants for the new roots grow just beneath the surface and radiate a good distance away from the plant and so are frequently damaged by hoeing too close.

Whenever the weather is dry, give the plants a thorough soaking. This is best done in the evening so that the water can percolate to the roots during the night. It is important to keep the roots moist so that cane growth is vigorous.

The first varieties will have ripened their fruit by about 1 July, when the first flush of strawberries are finishing. The best for freezing are those which part easily from the core, or plug, and so do not bruise or squash. These are Malling Notable, Malling Admiral and Norfolk Giant. Treat them like strawberries, placing them on trays, not quite touching, and place in the freezer for two hours. Then remove, place in plastic bags and consign to the freezer; they will retain their shape and keep for at least a year.

By planting for succession, there will be fruit until the end of autumn when the old canes (those that have fruited) are cut away 3 in. from the base and burnt. The new canes will have been tied in and will fruit next year but any damaged or weakly canes are best removed, leaving eight to ten strong ones to each root, though it must be said that some varieties do not make as much cane growth as others. Then give the plants a generous mulch and they will need no further attention until mulched again the following June. The culture is simple and most of the work is done without stooping.

To increase the stock, lift one or two roots (possibly where there is overcrowding) in autumn and gently pull away the canes with a portion of root and replant as soon as possible. Any part of the old root remaining is then discarded.

Always keep the canes away from drying winds for if the roots dry out they may not recover. Also, should canes arrive during a time when the ground is not suitable to work, heel them in or dig a trench for the roots and cover them with straw heaped over with soil. If the soil is frozen, place the canes in a dark shed or garage with the roots in a deep box and cover with straw or peat which must be kept moist. Plant as soon as the frost has left the soil and it is not too sticky to work.

Varieties

EARLY
Lloyd George. Until the arrival of the early Malling varieties, this was the first to mature and as it is rather

susceptible to frost, often fruited below its capacities. By 1950, it had become so troubled by disease that it was necessary to introduce a virus-free strain from New Zealand and once again it crops heavily, the bright-red coloured fruit being firm and of good size.

Malling Exploit. It ripens about a week later than Malling Promise and Lloyd George and is a hardy vigorous variety, bearing its scarlet berries in trusses, which are partly hidden by the foliage. The canes grow tall and branchy. In all respects it is hardier than other earlies and so does well in the north where it holds its fruit longer than any.

Malling Jewel. Follows Malling Exploit, ripening between the early and mid-season varieties and is the best all round variety yet introduced. It is frost hardy; in any case it blooms later than other earlies and is immune to virus. At the same time, it crops heavily, up to 6 tons an acre in a good year. It makes plenty of tall, smooth canes but only in a rich soil, and bears large conical fruits of deep crimson.

Malling Promise. The first of Mr Grubb's introductions and the first to ripen its fruit. The canes are borne in profusion and grow strong and upright, making tying-in an easy task. The blossom is quite tolerant of frost but in a severe winter the canes may be cut back and some fruit lost. Good for freezing and bottling, the berries are large, plug easily and are of a dull red colour.

Royal Scot. One of the best of all raspberries, it makes vigorous cane growth and is the first to mature in the north (Malling Promise being first in the south). The large salmon-red berries are borne all along the many branchlets and ripen over an extended season, making it a valuable kind for the home garden.

MID-SEASON

Delight. An early mid-season variety making plenty of

tall sturdy canes and bearing more heavily than any variety at the National Fruit Trials. Resistant to frost and virus, the large globular fruits possess outstanding flavour.

Glen Clova. A new Scottish variety which is cropping better than any variety in north Britain. It blooms late and so escapes the frosts; cane growth is vigorous and healthy. It bears bright-red fruits which retain their shape after freezing.

Malling Enterprise. One of the later mid-season kinds, it is frost tolerant but rather shy in making new canes unless grown in a heavy soil (for which it is the best variety) which is enriched with plenty of manure. The canes are smooth and the crimson fruit, which plugs easily, is large, sweet and juicy.

Malling Notable. Cane growth is not too vigorous and has the habit of bending over at the top, but this variety continues the season until the later Norfolk Giant is ready. The large berries, which plug easily, are of dull crimson and of good flavour.

Newburgh. Crops heavier on the western side of Britain where it makes plenty of cane growth, the berries being the largest of all, of bright crimson and plugging easily.

LATE

Malling Admiral. Possesses the same vigour and resistance to virus as its parent, Malling Promise, and is an equally heavy cropper. One of the last to ripen, its large fruits are amongst the best for freezing.

Norfolk Giant. The last of the summer kinds to ripen, it is a vigorous grower in all soils and is rarely troubled by frost or disease. The large red berries plug well and are of acid flavour thus being good for jam; they also freeze well.

Autumn-fruiting

These require the same culture as the others but fruit

in autumn on the new season's canes which are cut back after fruiting late in November. They continue the fruiting when Norfolk Giant finishes.

Varieties

Fallgold. A splendid American variety, making plenty of robust cane growth and bearing, until the end of autumn, large yellow berries of apricot flavour.

November Abundance. One of the few raspberries to receive an Award of Merit, it is the last to fruit and in warm southern gardens continuing until the end of November; its large crimson-red berries possess excellent flavour.

September. From America, it takes over when Norfolk Giant ends and fruits until late October though is best grown south of the Trent for it to do so; in northern gardens it may not ripen. It bears large red fruits of good flavour and is one of the best for a dry, sandy soil.

Zeva. A Swiss introduction and, as would be imagined, is extremely hardy. The fruits are larger than those of any other variety and ripen from early August until November. They are deep scarlet, sweet and juicy and freeze well. This variety makes plenty of cane growth each year.

Pests and diseases

Aphis. Besides attacking the leaves and tips of the shoots, causing the leaves to turn brown and fall, aphis (greenfly) are believed to cause the plants to be susceptible to virus disease which may lead to 'dwarfing', i.e. stunted growth of the canes. The disease enters at the points where the aphis have punctured the stems, causing swelling and ultimate loss of vigour. Control as routine by spraying the canes with tar-oil in winter and with Lindex or derris before the fruit sets.

Mosaic. The most troublesome of raspberry diseases causing the leaves to curl and bright yellow markings to appear on them. It is caused by a virus which lives in the sap and is possibly introduced by aphis. There is no cure so pull up any infected canes and burn them. Virus-free strains of Lloyd George from New Zealand and resistant varieties, such as Malling Promise and Norfolk Giant, should be grown where virus has proved troublesome.

Raspberry Beetle. An insect pest which also attacks the blackberry and loganberry, being the most troublesome pest of the raspberry, frequently seen on the ripe fruit. It is greyish-brown, about $\frac{1}{6}$th in. long and it lays its eggs in the flowers. After hatching, the white grubs eat the fruits and adhere to them after picking. To prevent an attack, dust the plants with derris as the flowers open and again when the fruit sets, as routine.

Raspberry Moth. This pest spends its winter in the soil near the canes as a cocoon then emerges in spring as a silver-brown moth with yellow wing spots. The moths lay their eggs on the flowers. After hatching, the caterpillars devour the young fruits so that they do not mature. Dust the flowers with derris as they open and as a further precaution, soak the soil along the rows in winter with tar-oil when the gooseberries are sprayed.

Cane Spot. (See page 114.)

7 Blackcurrants

General considerations
The blackcurrant, *Ribes nigrum*, with its aromatic leaves, is occasionally found growing wild in Britain and in other parts of Europe but it is not common and

unlike most of our other fruits which enjoy the cool climatic conditions of northern Britain, blackcurrants prefer the warm south. In the colder parts, they tend to drop their fruit buds, especially where prevailing winds are cold; they are also more troubled by frosts than most other fruits. Several recent introductions, however, show a marked tolerance to frost and this has given the blackcurrant a much needed new popularity in recent years. Also, there is now a wide range of varieties to fruit over a greatly extended period. The advent of Laxton's Giant has also given us the first real dessert currant and although black-currants have many uses, this one is worthy of growing for dessert alone.

Blackcurrants make a delicious conserve and a wine which, if taken at night, encourages sound sleep. The extraction of the juice, richer in vitamin C than any other fruit, is valuable in providing the body with this much needed vitamin in winter. Also, when taken hot the juice will ease a sore throat. The fruit also makes delectable tarts and flans so a place should be found for at least a few plants in every garden.

Blackcurrants enjoy a warm climate and crop better south of the Trent. If they are grown in the north or on exposed ground, give them some protection. Erect 6 ft high interwoven fencing against the prevailing wind, or plant a hedge of *Cupressus leylandii*, spacing the plants 18 in. apart. A 'hedge' of Himalaya Giant blackberries, trained against galvanised wires will also act as a reliable wind-break. Pollination of the flowers will not be as successful where the plants are exposed to cold winds and ample pollination is essential if the plants are to set heavy crops.

Soil requirements

Blackcurrants crop well only in a heavy soil but it must be well drained in winter. If it is not, dig in anything that can be obtained to improve drainage. Clinker of boiler ash, crushed bricks obtainable from a building site, shingle or clearings from ditches are all suitable. Then incorporate some peat and decayed manure, for blackcurrants must have plenty of humus, to keep the ground moist in summer, and nitrogen for the plants to make plenty of new wood each year. No fruit requires more nitrogen in its diet and only one variety, the old Baldwin, does really well in a light soil. Composed straw, shoddy, poultry manure or garden compost should be worked in and where only limited supplies are available, give the plants a 2 oz per square yard dressing of sulphate of ammonia at the end of March. Light land, especially, will need all the humus it can be given if it is to grow good crops and when the plants are established, in their second year, give them a mulch of strawy manure every March and again in November when they are pruned. In addition to nitrogen, black-currants need constant supplies of moisture through-out the summer to make new wood and to form sizable berries which will then be sweet and juicy. Also, plant only in clean ground. Soil that is infested with weeds and couch grass will have been deprived of much of the nitrogen that should go to the currants.

One is often asked how much fruit may be expected from a plant. With blackcurrants, so much depends upon soil and situation whilst every season is different as to late frosts and cold winds. But in an average year, well-grown plants will yield, when four to five years old, 6–7 lb of fruit. Where growing on a commercial scale in Somerset, Wellington XXX usually gave that amount of fruit and the amateur who can give the plants individual attention may expect considerably more, but with this fruit more than any other it may be said 'there are horses for courses'. Some varieties do well in certain parts of Britain, others not so well.

It will be necessary to try several and then to increase those most successful, by cuttings which root easily.

As with other soft fruits, it is essential to obtain clean plants—those guaranteed free from 'reversion' and 'big bug' (caused by Gall Mite) which trouble blackcurrants in the same way that virus diseases trouble strawberries and raspberries. Obtain your plants from specialist growers who will supply only the best, for remember, as with gooseberries, the plants will bear heavily for at least twenty years. This is also why it is important to plant in clean, well-manured ground.

Planting

Two-year bushes are the best to plant and as they root easily, they are inexpensive. Perhaps two bushes of each of four varieties will give worthwhile pickings, planting for succession, not only to spread the crop from early July until early October but so that if late frosts spoil the blossom of the early varieties, there will be others to follow.

Plant any time between 1 November and early March. If the soil has a high clay content, it is better to plant in March, for with warmer days the roots will begin activity and will not lie dormant as they would if planted in November when excessive rain during winter may cause them to perish. Do not plant when the ground is wet or contains frost. It should be in a friable condition so that the plants can be trodden in.

The more vigorous kinds, e.g. Wellington XXX and Mendip Cross, should be planted 6 ft apart each way and those of more compact habit, about 5 ft apart. For the small garden, plant those of upright habit, such as Seabrook's Black and Amos Black, with Laxton's Giant for dessert.

An open, sunny position is required, for unlike gooseberries in so many ways and in this respect

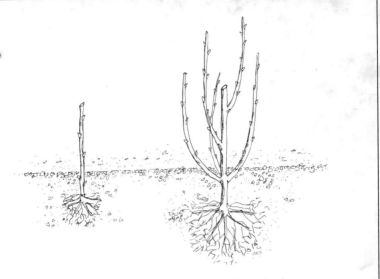

Fig. 10 Blackcurrant cutting taking root (left) and new bush (right).

especially, blackcurrants will not do well in semi-shade. They also prefer deeper planting than other fruits. See that the roots are spread out and well covered with soil and the basal buds are just below soil level. Before covering the roots, give them a handful of bone meal, which will release its nitrogen content over a long period.

Care after planting
In March, cut back all shoots to about 3 in. of the base, as with young rose trees. This is to encourage the plants to produce more basal shoots which will develop to carry the crop the following year. There will be no fruit the first year but the aim is to build up a strong plant. For the next two years, no pruning will be necessary. Afterwards, cut back older wood so as to maintain a balance between the old and new for the crop will be carried on all the wood, hence heavy crops are obtained from established bushes.

89

November is the best time to look over the plants, removing any dead wood or branches broken by wind. Overcrowded shoots are cut out near the base and where the shoots may be excessively long, cut back to a bud from which they will 'break' into new growth. The idea is to build up and maintain a healthy, well-shaped bush so that the sun can reach all parts.

When the plants are young, remember to look over them at the end of each winter and to tread around them if made loose by wind.

Blackcurrants do not grow on a 'leg' like gooseberries (although red currants do) but develop from base buds. To increase the stock, young shoots are removed about 3 in. above soil level and shortened to about 12 in., retaining the buds all the way along the stem. From these buds the plants will send up new shoots.

Cuttings are taken early October and inserted in trenches of sand and peat. Make a V-trench 6 in. deep and insert the cuttings to a depth of 3–4 in. They root without difficulty but will do so more quickly if the end is treated with hormone powder before planting. The cuttings are planted 4 in. apart and made firm. During summer mulch the rows and water well in dry weather. By October, a year after planting, they will have rooted. They are lifted carefully and replanted at the correct distance apart in prepared soil. The following March, cut back the shoots to 3 in. then the plants will make a good-sized bush to come into fruit the following year. Strawberries can be grown between the rows of young plants so as to make the best use of the ground until the currants have grown bushy.

Do not pick the fruit too soon; wait until it has turned jet black. By then it will have made some size and will be juicy and full of flavour. But do not allow the berries to become over-ripe and do not pick when the fruit is wet after rain. First, let the fruit dry and, when picking, hold up the strings (as the stems are called) and remove the berries with care so as not to bruise them.

To freeze, place in cellophane bags with care and, after closing them, place in the deep freeze without delay. The berries should have had any stalks removed before placing in the bags and in this respect take longer to prepare than other fruits, with the exception of strawberries.

Varieties

EARLY

Boskoop Giant. Introduced from Holland in 1895, it forms a large bush and a long fruit truss, the berries being of top quality. Flowering early, it is not as tolerant of frost or cold winds as others and grows best in a warm, sheltered garden.

Laxton's Giant. The first and only dessert variety, the fruit of a well-grown plant being as large as an Early Rivers cherry. It is eaten like a gooseberry (or cherry) and is sweet and juicy. It bottles well and makes a delicious flan or tart. Maturing by 1 July, it bears heavily and is tolerant of frosts and cold winds but requires a heavy well-nourished soil if the berries are to reach maximum size and sweetness. It has won more awards for exhibitors than any variety and it has the ability to retain its fruit in condition for several weeks.

Mendip Cross. One of the best earlies for a cold district, it has all the good qualities of its parents— Baldwin and Boskoop—and none of their faults. It is tolerant of frosts and cold winds and bears a heavy crop of juicy fruits of medium size.

Tor Cross. A new introduction which bears heavy crops of large, thin-skinned fruits and does not readily drop its buds in cold winds. It makes a compact plant and crops well in light soils.

Wellington XXX. Raised by Capt. Wellington at East Malling, it makes a large spreading bush and is the best all-round variety ever raised. It follows Mendip Cross but blooms later and so escapes the frosts. As it does well in most soils though requires plenty of nitrogen, it is one to grow in every garden. Its thick-skinned fruit freezes better than any.

MID-SEASON

Blacksmith. Raised by Thomas Laxton, it is with Baldwin, one of the few to crop well in light soils. It blooms late and is frost proof. It bears its large fruits in double trusses making it one of the heaviest cropping varieties. The berries are noted for their high vitamin C content.

Seabrook's Black. This follows Wellington XXX and makes a compact, upright bush, ideal to plant where space is limited. Though ripening quite early, it blooms late and so misses the frosts. It is also resistant to 'big bud'. It bears a large, rather acid berry which is the best for jam and freezes well.

Westwick Triumph. Probably the best to follow Wellington XXX. Though vigorous, it is of upright habit, and flowering late, it misses the frosts so bears consistently heavy crops. The fruit is large and borne in long generous trusses.

LATE

Amos Black. A Baldwin cross raised at East Malling, it has the same compact upright habit as its parent and so is valuable for small garden planting. It blooms later than any, thus escaping all frost, and also ripens its fruit later than any, at its best in early October. The berries are medium size and have a thick skin, so they freeze well.

Baldwin. The Hilltop strain is the best but like Leveller gooseberry it crops well only where soil and climate suit it. It does best in the south and requires a light soil but one which contains plenty of nitrogen to form new wood for it makes a small plant. The berries are large and it holds its fruit well when ripe.

Cotswold Cross. It has Baldwin as a parent and crops at about the same time. Though raised at Long Ashton, Bristol, it also crops well in the eastern half of England, the berries being large and borne in short dense trusses.

Laleham Beauty. Raised by Mr R. Salter, it has received an award from the Royal Horticultural Society. With Amos Black, it is the latest, holding its fruit until early October. It crops heavily in most soils. The berries are large and thick-skinned so making it valuable for freezing.

Worcester berry

A natural hybrid of the blackcurrant and gooseberry which has the characteristics of both parents. It grows wild in North America.

In a soil enriched with humus-forming manures, it makes a bush 5–6 ft tall and almost as bushy. Its sturdy arching branches have thorns but as it bears its fruit in long sprigs or trusses, like blackcurrants, it is easy to pick. It makes an excellent hedge and may be planted as a wind-break for it does not possess the blackcurrant's hate of cold winds. Plant 4–5 ft apart and in a well-fortified soil, treading in the plants to prevent wind movement.

It is ready to pick in late summer, the fruit combining the characteristics of both parents, the berries being crimson-black, of the size of Laxton's Giant currant but with the true gooseberry flavour. Fruiting over several weeks, it is excellent to stew or to use in tarts and flans; it also makes a fine preserve.

Pests and diseases

Gall Mite. This is the pest which causes 'big bud', whereby the buds swell with the pests inside them and produce no fruit. On plants that are attacked, the mites are present in their thousands though they are only one hundredth of an inch long and can be seen only through a powerful microscope. Control is difficult, but as routine, although blackcurrants are sulphur-shy, they will take a one part in fifty lime-sulphur solution which, if used each spring, will usually be sufficient to keep the plants clean. Wellington XXX, Blacksmith and Westwick Triumph are especially sulphur-shy and the application should be further diluted to one part in sixty or seventy when used on them.

Green Capsid Bug. Often present on old plantations, feeding on the leaves and buds and causing loss of crop. The pest can easily be seen and can be controlled by spraying with a mixture of tar-oil and petroleum in early spring, before the eggs are laid in April.

Leaf Spot. Attacks the plants as brown spots on the leaves and wood and if not controlled will reduce the vigour of the plant. Spray with Bordeaux Mixture, as for gooseberries, or with Orthocide, either of which should be applied after the fruit has been gathered.

Reversion. So destructive is it that only stock certified as being clear of it should be obtained. It is caused by a virus carried by a mite and is usually noticed (if present) in June when the young leaves are seen to have grown long and narrow with smooth edges. Soon they turn dark green and fall, greatly reducing the vitality of the plant. Spraying with a weak solution of lime-sulphur (one part in fifty of water) as routine will do much to prevent an attack but there is no cure once it appears and the plants should be dug up and burnt. It is important to begin with clean stock.

Rust. A troublesome disease recognised by brilliant orange spots on the underside of the leaves. It is also present on pine tree needles and for this reason blackcurrants should not be planted near pinewoods. The leaves will fall prematurely but the trouble may be controlled by spraying with Bordeaux Mixture early summer and again after the fruit is gathered.

3 Red and White Currants

Soil requirements

Ribes rubrum is the red currant and *R. album* is a white form of it. At one time both varieties were widely grown for their high pectin content (which is used in the setting of jam) but, nowadays, pectin is extracted mostly from apples. However, red and white currants make a delicious jelly to serve with mutton and cold meats and with raspberries make a pleasant 'sweet', especially if served with cream or a sprinkling of red wine.

These currants are like gooseberries in that they grow on a 'leg' and require much the same culture. But they share the same enjoyment as the blackcurrant in their liking for a heavy soil. Growing on a 'leg', they are not so heavy cropping as blackcurrants for they do not make big plants. They are not troubled by frosts but hate cold winds. More than any other soft fruits, however, the red currant suffers from birds taking the berries when they have turned brilliant red. Blackbirds and bullfinches are the chief culprits and as the berries begin to colour, the plants should be covered with muslin or netting. This will not prevent the berries ripening. To delay for a single day may mean losing the entire crop.

Culture

Always shy in producing new wood, these currants need a soil containing plenty of nitrogen, so work in as much decayed manure as possible. Used hops or shoddy are also suitable; or use straw composted with an activator and to which has been added some poultry manure. In addition, peat is useful to provide humus in the soil.

Like gooseberries, these currants need some potash, so in spring each year, sprinkle around each plant 1 oz of sulphate of potash. This will increase the quality of fruit and help it to colour.

It is also essential to plant into clean ground for it is not possible to clean the ground afterwards without damaging the surface roots. A well-drained soil is also necessary so that winter moisture can quickly drain away. Humus is, however, needed to retain summer moisture without which the berries remain small and seedy.

Planting may take place at any time between 1 November and mid-March when the soil is in a friable condition, but November planting, when the soil is still warm, will enable the plants to become established before the frosts. Before planting, any roots must be removed which have formed up the 'leg' or they will produce 'suckers' which sap the energies of the plant. This treatment is also necessary for gooseberries. Remove these roots with a sharp knife.

For red and white currants, select a place sheltered from strong winds which could cause the rather brittle wood to break, for the plants make only a small amount of wood. Where growing as bushes, they may be set out about 4 ft apart each way, though Laxton's No. 1 should be allowed an extra 6 in. Tread around

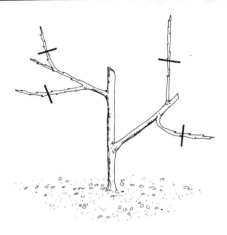

Fig. 11 Pruning a red currant.

the plants firmly and in March cut back the shoots to 3 in. from their base to a healthy bud. This will encourage the shoots to break into new growth and a well-balanced head can be formed. Purchase two-year plants and though there will be no fruit the first year after planting, there will be some the following year and the amount will increase each year.

These currants should be given a generous mulch in early June and again in October when the plants are looked over and any dead or damaged wood removed. Otherwise their pruning requirements are at a minimum. For the mulch, use peat mixed with decayed manure or composted straw or used hops.

During dry weather, from the beginning of summer, give the plants plenty of water and every fourteen days, an application of dilute manure water. This will enhance the size and flavour of the berries. No other soft fruit (except the gooseberry) responds better to this feeding.

Red and white currants can just as easily be grown as single or double cordons, like gooseberries, and may be grown against a sunny wall or fence, or in rows, fastened to strong wires. They will grow to 5 ft tall and make the best use of limited space. Where growing in this way, the fruit can be protected by draping fine mesh netting over the plants from hooks fixed into the fence or wall. Single cordons are planted about 2 ft apart, and double at 4 ft.

A single cordon is formed from a rooted cutting. When rooted, cut back to a bud about 6 in. above ground and grow on the extension or leader shoot to the required height, each year, pinching back the side shoots to about 3 in. of the stem. The double cordon is formed by cutting back to a similar height, but to two buds, one on either side of the stem. The shoots from these buds are grown on, first at an angle of 45°, tying them to canes, then gradually reducing them to an angle of 90°. These in turn are cut back to upward pointing buds and the shoots grown on vertically, pinching back the side shoots to 3 in. of the main stem.

Cordons and bushes are formed by taking cuttings from established plants, using the new season's wood which is not too hard. The best time to take these cuttings is October. The cuttings should be about 15 in. long and all except the four upper buds are removed so that the plants will form a 'leg'. Treat the end which is to form roots with hormone powder for these currants are as difficult as gooseberries to root. Then plant them in 6 in. deep V-trenches which have been filled with a peat and sand mixture. Plant firmly, spacing the cuttings 3 in. apart. Leave them in the rows for a year, keeping them well watered in summer, then move them to their fruiting quarters. Whilst rooting, give the rows a mulch in June to prevent moisture evaporation. With established plants, the first fruits will begin to colour early July and they should be carefully watched so that they can be gathered at the right moment. If allowed to hang too long the fruit will be soft and little good for any purpose, that is, if the birds have not taken them by then. The fruit will freeze well only if the berries have not burst their skin.

Varieties—red currants

EARLY
Fay's Prolific. An American variety, as most of the best red currants are. As bud burst is late, it misses the frosts and is the best early for northern gardens. Though making a small bush, it crops well, the crimson fruit being sweet and juicy.

Jonkheer van Tets. A Dutch variety of recent introduction which makes a strong plant and crops heavily, the berries being large and of good flavour.

Laxton's No. 1. A report of the National Fruit Trials recommended it as the 'outstanding red currant for commercial purposes'. It is the earliest to ripen and is a strong grower, cropping well in all parts.

MID-SEASON
Houghton Castle. Raised near Hexham as long ago as 1820; it is hardy and flowers late so escaping the frosts. A vigorous upright grower, it crops well in most soils, the medium-sized fruits ripening to dark, dull red.

Red Lake. Probably the best red currant and if there is room for only a few plants, they should be of this variety. Raised at New York State Agricultural Station, it makes a large upright bush and bears heavy crops of enormous berries, spaced evenly about the trusses. The fruit is brilliant red with crimson veins and is the exhibitors' favourite. It withstands wet weather well.

LATE
Laxton's Perfection. When introduced in 1910, it caused a sensation with its large bright-red fruits

which ripen late in August, thus prolonging the season. Growth is strong and vigorous and it crops heavily in well-nourished soils.

Wilson's Long Bunch. The last to mature, it is a hardy variety, its medium-sized fruits being produced with freedom and held in a long narrow bunch. They ripen to deep pink and have a rather acid taste but are excellent for preserves.

Varieties—white currants

This fruit has its own special flavour and is used to mix with red currants, either for flans or jam making, as used on its own, it turns yellow after cooking. It is also served uncooked with red currants, providing a most attractive 'sweet' dish. The best varieties are:

White Dutch. It makes a small, spreading bush and follows White Versailles in its flowering and fruiting. The milky-white fruits are borne in 3 in. long bunches and are richly flavoured when fully ripe.

White Transparent. The latest white to extend the season and the exhibitors' favourite. It blooms late, the bunches being long and the yellow-skinned fruit of large size but acid to eat raw.

White Versailles. Raised at Versailles in 1843, it is the best of the whites and the earliest, the good-sized fruits being pale yellow and borne in large trusses 4 in. long. It is sweet and juicy. It makes an upright bush of moderate vigour.

Pests

Clearwing Moth. Mostly confines its attacks to red and white currants and will lay its eggs on the wood in summer. Upon hatching, the grubs tunnel into the stems, feeding on the sap and causing the branches to die back. As routine, spray with tar-oil in winter as for other fruits. When once the grubs enter the plants, control is difficult.

9 Blueberries and Cranberries

Blueberries—new species of value

Present on peaty moorlands throughout the British Isles, being especially prolific on the Yorkshire Moors, on Dartmoor and in Scotland and Ireland, the blueberry is also known as the bilberry and whortleberry. It is a hardy shrub-like plant with small ovate leaves and bears multitudes of dark-blue berries covered with a grape-like 'bloom'. They are, in fact, like small black grapes and are ripe from mid-August until early November when they are much in demand for tarts and flans and for making into a delicious conserve. In the south-west they are eaten raw, served in small fruit dishes with clotted cream.

The plants are also present in North America and across northern Europe and Asia, growing about 12 in. tall and always in large colonies. Introduced to gardens from the wild, they never crop well and it was not until Dr Colville of the US Department of Agriculture introduced *Vaccinium corymbosum*, with its striking foliage of bronze, crimson and gold in autumn (worthy of planting in gardens for this feature alone) and bearing much larger fruits than the European species, that the blueberry was considered worthy of commercial planting; its fruits are at least twice as large as those growing wild.

Soil requirements

The blueberry is a splendid fruit for those gardens of an acid soil, i.e. with a pH value of about 4.5–5.0, and requires the same conditions as the azalea and rhododendron. If the soil has a neutral reaction and does not have a chalky subsoil, the plants may still be grown provided plenty of low grade (more acid) peat is worked in. Peat of this quality is inexpensive to buy or it may be collected from moorlands for the taking, two or three sacks being sufficient for about twelve plants.

The plants require plenty of moisture at their roots and grow well in land that is not too well drained. Where the soil is sandy, work in some humus, such as decayed manure or composted straw, together with plenty of peat and, at the same time, clear the ground of weeds. Shoddy and poultry manure are also suitable, for the plants require plenty of nitrogen to make new wood. Where this is not available, the plants will eventually become a mass of old twiggy wood resulting in reduced cropping.

In a soil containing plenty of nitrogenous manure, the plants will eventually grow 4–5 ft tall and make large bushes, each capable of producing 14 lb of fruit when four to five years old. They make a handsome sight when in flower in early summer; the long elegant sprays are clothed in bell-shaped flowers of palest pink, puckered at the mouth, like lily-of-the-valley. They remain in bloom for several weeks and are followed by the first fruits of the early varieties which ripen about mid-August.

The first to ripen are Early Blue and Pemberton, followed by Rubel and lastly Jersey, which should be confined to southern gardens if it is to ripen all its fruit before the frosts. The plants may be grown as game cover on large estates or on an unsightly bank or beneath young trees requiring similar soil conditions—blueberries are tolerant of partial shade but not of cold winds. They enjoy a moist climate and plenty of moisture about their roots in summer. For this reason they do better in the west where the winds are less cold and drying. If growing in exposed gardens, provide the plants with some protection against cold winds, planting a blackberry hedge as a wind-break.

Planting

March is the best time to plant and if the land is low lying and damp, the plants will settle in whilst conditions are favourable. But on well-drained land, planting may take place at any time between November and March whenever the soil is in a friable condition.

Deeper planting is necessary than for other soft fruits and this will encourage the plants to form numerous sucker-like shoots from below soil level. Plant 2–3 in. below soil level and tread in firmly. The plants will grow bushy and should be set 4–5 ft apart. For game cover, allow 5–6 ft. Too close planting will deprive the centre of the plants of sunlight and air with the result that they will form a lot of brittle wood which will gradually decay and die back.

To help with pollination, it is advisable to plant at least two varieties together. The American growers who cultivate this crop on a commercial scale, feed the plants when three years old and each year after with a balanced diet in addition to a supply of nitrogen. This consists primarily of 2 oz per plant of sulphate of potash and superphosphate of lime in equal amounts. They find this greatly increases the yield and they often obtain as much as 20 lb of fruit from each plant when four to five years old. Given good cultivations, they will crop heavily for at least twenty years before beginning to fall off through forming an excess of old wood.

Plants are expensive, costing about £1 each, but are reasonable at that when taking into consideration their long life and the amount of fruit they produce.

The first fruits will be ripe by mid-August and the

plants must be picked over at least once a week for about eight weeks. Later varieties will extend the season for another four weeks. The season coincides with blackberry picking, the two crops yielding fresh fruit when all other fruits, apart from autumn strawberries, have finished. Blueberries must be picked with care for they bruise easily. Marketed in 1 lb and ½ lb waxed punnets, the fruit finds a ready sale and is not to be compared to the wild bilberries marketed in large baskets, the fruit often in a squashed and over-ripe condition. The berries will have turned from green to red, then to pale blue before being fully ripe when they are almost black and covered with a pale blue 'bloom'. The berries must not be picked until quite ripe otherwise they will be devoid of flavour. They are borne on short sprays of ten or twelve along the slender shoots and are held well above ground, being as easy to pick as black-currants. To protect the fruit from birds (and game birds have a liking for the berries) it may be necessary to cover the bushes with muslin or netting from early August, possibly using that which has been removed from the red currants or strawberries. This will not hinder the ripening of the fruits.

Care after planting

As the plants must have plenty of moisture to make new wood and to increase the size of the berries as they mature, a mulch should be given in June, right up to the plants. This may consist of decayed manure or straw composted with an activator and some poultry manure, which will supply the plants with additional nitrogen as well as preventing moisture escaping from the soil in dry weather. It will also suppress weeds, which use up moisture needed by the plants, and will reduce cultivations to a minimum. Peat is also excellent for mulching this crop and should be used in quantity before applying the manure. The plants will benefit from a further mulch given in November after any pruning has been done. If hoeing the ground at this time, take care not to work too near the plants for fear of damaging the surface roots.

This is a good crop to plant on scrub-land which may have been cleared of forest trees and which may be situated some distance from home. It will be necessary to fence off the land from trespassers, and from rabbits, sheep or deer which will nibble the plants. To guard against rabbits, wire netting must be fixed to the posts at least 10 in. below soil level, and exclude man and animals to a height of 5 ft.

Propagation
As the plants are expensive, it will be necessary to propagate one's own but as this plant will increase by more ways than any other soft fruit, it is difficult to understand why the plants are so expensive.

Cuttings (shoots) 6 in. long of the new season's wood may be removed during August and after treating them with hormone powder, inserted into a mixture of sand and peat used to fill a V-shaped trench made 6 in. deep. Plant them 3 in. apart and firmly. They will have rooted by the following autumn when they should be carefully lifted and planted where they are to fruit. If the trenches are covered with cloches, the cuttings will root more quickly and much time will be saved. The cuttings must be kept moist in summer.

A simpler way is to 'tease' away the suckers, which form around the plants, with their roots attached. These may be planted at once into their permanent quarters, spacing them 2–3 ft apart and removing alternate plants after three or four years and replanting.

Again, shoots can be layered, bending them over until they reach the soil, then partially severing them by making an incision with a sharp knife and bending back the cut part so that it comes into contact with the soil. Hold this in place with a strong wire pin or wooden peg, exactly as when layering a carnation. The best time to do this is July. Keep moist the point of contact with the soil and the shoot will have begun to form roots from the point of incision by the following April. The shoot is then removed from the parent and replanted.

Blueberries will also grow well from seed though will take two years longer to come into fruit than when propagated by vegetative methods. Seed is sown when ripe in October, in shallow drills in a soil made friable by the addition of peat and cleared of stones and weeds. Make the drills 1 in. deep and sow thinly. Germination will take place by early the following summer. The seedlings must be kept moist during dry weather and given a peat mulch in June. By early autumn they may be moved to beds containing plenty of peat and grown on for another year before they are ready to move to their fruiting quarters. They will produce good crops though the quality of fruit will not be as good as from named varieties.

Cranberries

Cranberries are not nearly so common in the wild in the British Isles as they are in North America, nor so often cultivated here, yet they are excellent plants for rough land where the soil is of an acid nature. Hardiest of all fruits, they flourish in Canada, Newfoundland, Alaska and across northern Europe, where *Oxycoccus palustris*, to give it its botanical name, grows in marshy bogland. It is a small-leaved twiggy shrub growing 2 ft tall, similar to bilberries in the wild, and the fruits, which ripen in September and

October, are brilliant crimson, like big red currants and with a similar sharp taste. In America, they are used to make tarts and flans and a preserve or sauce to accompany roast turkey.

Forming long wiry stems, they are plants of almost creeping habit and if planted 2–3 ft apart will quickly form a dense mat completely covering the ground so that no weeds can grow beneath. Plant in autumn or spring, preferably in low-lying land, for the secret of success with this plant is that its roots be submerged for weeks on end, through summer if possible. Leave the hose on the plants for a day or night each week if the summer is dry, to give them a thorough soaking. The plants do better if the ground is given a thick layer of peat and sand after planting to act as a mulch.

Propagation is by lifting and dividing the roots in winter into small pieces, as for heathers; and by layering the shoots as for bilberries.

Picking begins mid-September and continues until early November, depending upon frosts; these will make the fruit soft and impossible to pick. By early November, the fruit will have lost its flavour and picking should be abandoned.

10 Blackberries, Loganberries and the Hybrid Berries

Blackberry species and varieties

The blackberry or bramble, *Rubus fruticosus*, is a native of all parts of the British Isles; the loganberry is native to North America. Both require the very opposite conditions to be successful. North America is also the home of a wide range of blackberry species, notably *R. villosus* and the wonderful Oregon Thornless, one of the cut- or fern-leaved blackberries. In Britain there are few species in comparison to those found across northern Europe and Asia but a number of sub-species are present such as the Dewberry and Hazel-leaf bramble and one that is indigenous only to the Warwickshire woodlands and hedgerows, having a more upright habit and with canes more like those of the raspberry but branching from near the top, about 5 ft above ground level and then rambling on for 12 ft or more. The black fruits are large and of splendid flavour. Then there is the Parsley-leaf bramble, a seedling of the cut-leaf bramble, *R. laciniatus*, a handsome plant which, with Oregon Thornless, may be used to cover an arch or trellis, possibly dividing one part of the garden from another. These plants are so decorative in leaf that they are worth planting for this purpose alone. But, in addition, they give large pickings of fruit from early August (when the early varieties mature) until the end of October. Each of these species and sub-species has been used in the raising of numerous varieties which now prolong the season, but perhaps that which was to revolutionise blackberry growing in Britain more than any other, came to us from the Himalayas where seed was obtained by the German botanist, Theodor Reimer, shortly after World War I. The late Edward Bunyard thought this a 'con' trick, that the new species —Himalaya Giant—was but a seedling from *R. pubescens* or *R. thrysanthus*, both being found in North America and northern Europe. For all that, its hardiness, enormous vigour and tremendous cropping, each shoot carrying up to fifty shining black berries, brought it immediate popularity. It may be planted as a wind-break, the shoots being trained along strong galvanised wires, and as such will make an impenetrable hedge. If the shoots are kept under control, it may be planted for this purpose and, in addition, will give large quantities of fruit every autumn.

The blackberry is of great hardiness and is also untroubled by wet weather, the rain falls from the glossy berries, whereas with loganberries (and raspberries) water tends to stick. Blackberries are therefore ideal plants for the western side of Britain; also for planting up to 1000 ft above sea level. In the bleakest parts, grow them to fruit in autumn (planting for succession) with the equally hardy rhubarb for spring use and gooseberries (also for succession) for the summer months. These fruits are amongst the best for freezing and where this is possible there will be no wastage of the surplus fruit. They also make excellent jam. Again, there is always a demand for the fruit if the surplus cannot be frozen, marketed in 1 lb and $\frac{1}{2}$ lb punnets and sold to local food shops. Blackberries grown to the highest standards are the equal of any dessert fruits. The berries of Himalaya Giant often measure 1 in. across, so may be enjoyed raw as dessert or as a 'sweet' served with cream or sprinkled with claret. The berries of several varieties can be enjoyed in the same way for if grown well, they are sweet and juicy, the seeds almost unnoticeable.

Where to grow them

Blackberries may be grown in rows and trained along wires, or against a trellis or rustic fence, and they are happy in semi-shade, being like gooseberries in this respect. Again, they may be grown against strong posts which are about 6 ft above ground, tying in the shoots as for rambler roses. In this way, they take up little space and may be planted at the back of a shrub border or in the kitchen garden, planting them 6 ft apart. If the thorns of some prove unpleasant, plant Oregon Thornless instead.

Where training the shoots along wires and in rows, fix the wires to strong stakes driven well into the ground and about 5 ft above soil level. Here, plants of Himalaya Giant are set 10 ft apart and they will soon grow into each other. Less vigorous types are planted 6–8 ft apart, allowing the same distance between the rows. For those who wish to know the weight of fruit to be expected, from the plantation on my Somerset fruit farm in the 1950s, about 7000 lb an acre was the average, i.e. 7–8 lb per plant. But for the amateur's garden where the plants can be given greater attention in their culture, this weight can easily be doubled, at least 14 lb per plant being a fair average for established plants.

Plants cost about 80p each but it should be remembered that with good culture they will remain healthy and vigorous for thirty years or more, so that, as with gooseberries, the initial planting will repay its outlay many times over. In addition, the plants are easily propagated and whilst three or four will give worthwhile pickings for a start, these may soon be increased to a dozen or more by rooting the tips of the shoots.

Soil requirements

Blackberries and the hybrid berries require a soil containing plenty of moisture-holding humus if the berries are not to be small and seedy. The plants enjoy a stiff loam, well drained in winter but one which retains its moisture in summer. They will not do well in water-logged ground nor in a thin subsoil over chalk but will flourish in the acid soil of moorland peat, where blueberries also grow well. Nor do they grow well in sandy soil, the canes remaining small and weak, the fruit also being small. If the soil is light, work in plenty of humus. This may be garden compost or straw composted with poultry manure, shoddy, brewer's hops, or decayed farmyard manure; anything to give the soil 'body'. If of a high clay content, work in anything to help drainage, such as clearings from ditches, gravel or shingle. Peat is also excellent and should be used in quantity for all soils in which blackberries are to be grown.

Plant at any time between November and March when the soil is friable. Set the roots just below the surface and spread them out well. Cover with peat before filling in with soil and tread in firmly.

In April, cut back the canes to 3 in. of the base and at this time give the plants a mulch of peat and decayed manure or garden compost. Early July, retain and tie in the two strongest of the newly formed canes, removing any others. At the end of autumn, shorten them to about 5 ft for it is on these canes that the first fruit will be produced the following year. In July, pinch back the laterals to the third or fourth joint to encourage the plants to concentrate their energies on the formation of large juicy fruits. After fruiting, cut out at soil level any old canes which have begun to die back to encourage the new season's canes to bear the bulk of the crop. The plants must not be expected to carry too much old wood or the weight of fruit will become less and less instead of increasing each year. It is also advisable to remove any weak new canes when pruning is done. After pruning, give a generous mulch.

This process of pruning and shortening the canes is done every year, so keeping the plants free of dead wood. Also, provide them with a balanced diet—nitrogen to make new canes each year and phosphates to stimulate root action; bone meal which supplies both, suits them well. A handful to each plant should be given in autumn, raking it lightly into the surface.

The plants will also respond with large juicy fruits if fed occasionally in summer with dilute liquid manure or dried blood.

If the plants are slow to make new cane growth during a cold spring and early summer, give the ground a light dressing (1 oz per yard) of nitrate of soda, preferably in showery weather. This will stimulate the plants into growth.

Do not cultivate too close to the plants or the surface roots may be damaged. Mulching will suppress weeds and provide these roots with food and moisture.

Propagating

Propagation is simple. In July, the tips of those shoots needed for rooting are pressed into the soil near the plant and held down with a clothes peg or a metal layering pin. If the soil lacks peat, work in a little where the shoot is to be layered and keep the ground moist through summer. By early spring, the tips will have formed plenty of roots. The new plants are then severed from the parent with about 4 in. of cane and re-planted into their permanent quarters. During summer, new canes will arise and the two strongest are tied in, any others being removed. The canes will fruit the following autumn, two years after the tips were inserted in the ground.

To prune those growing against a post, cut away the twine used for tying in the canes and lay the canes on the ground around the post. Then remove any dead or weak wood and tie in the healthy young canes which remain.

Varieties

EARLY
Bedford Giant. Though not cropping so heavily as

Himalaya Giant, it is the favourite of the canners and for freezing. The fruit is large and sweet and retains its firmness better than any after freezing. It is also the first to ripen, often before the end of July.

Merton Early. Raised at the John Innes Institute, it ripens early, by late July or early August. It bears heavy crops of large juicy fruits of exceptional flavour. This blackberry comes true from seed which is sown in a frame early April, the seedlings being transplanted to beds of prepared soil in July where they are grown on until the following March when they are moved to their fruiting quarters. Of compact habit, plant 5–6 ft apart for it makes short canes which die back after fruiting and must be cut out. Next year's fruit is borne on the new canes.

MID-SEASON

Ashton Cross. Raised at the Long Ashton Research Station in 1937, it is hardy and shows marked resistance to virus. It crops heavily but is not as reliable as some. It is an early mid-season variety, bearing large, round black fruit.

Himalaya Giant. The most vigorous of all, the canes grow to 12 ft long and as it fruits as well on the old wood as on the new, it needs little pruning, merely cutting out any dead wood. Each cluster carries as many as fifty large jet-black berries which are sweet and juicy.

Oregon Thornless. A parsley-leaf blackberry with handsome fern-like foliage and as it is without thorns, the fruit is easily gathered. The canes grow to about 6 ft and it should be planted this distance apart. The fruit is of good size and borne in large heavy trusses. It is sweet and juicy and ripe by early September.

LATE

John Innes. Not ripe until October, it should be grown

south of the Thames otherwise in most years it will not ripen at all. Raised by Sir Daniel Hall in 1923, it extends the season until November and fruits well on the old wood, thus cropping heavily. The fruits are sweet and of good flavour.

Pests and diseases

Aphis. Attacks the blackberry in the same way as it does the gooseberry (see page 73).

Cane Spot. In addition to the blackberry, this disease also attacks the raspberry, causing purple spots to appear on the canes and foliage when they will die back. To control, spray with Bordeaux Mixture (see gooseberry) in early summer, before fruit set, not afterwards.

Raspberry Beetle. It also attacks the blackberry but not nearly so much. For treatment, see page 85.

Loganberry

This excellent but much neglected fruit, with its 1 in. long crimson berries, was originally found by Judge Logan who discovered it growing in California amongst a thicket of *Rubus vitifolius* of which it is thought to be a red variety. It was introduced into Britain by Bunyards of Maidstone in 1897 and as it does not part from its 'plug' when gathered, immediately became popular for bottling and canning. It also freezes well, keeping its shape, but must be allowed to become fully ripe before picking, otherwise it will be acid and devoid of flavour.

The plant fruits only on the new canes, like raspberries, and so that it will form plenty of new wood each year, it requires a diet containing nitrogen. This will also enable the fruits to grow large and juicy. So dig in plenty of composted straw, poultry manure or decayed farmyard manure and where these are not obtainable, add a handful of bone meal to the soil before planting into clean ground.

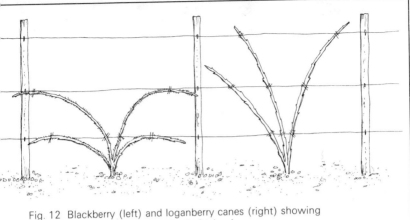

Fig. 12 Blackberry (left) and loganberry canes (right) showing different habit of each and method of tying in.

Loganberries produce their canes in a more upright fashion than blackberries and are tied to the wires fan-wise; blackberries are tied horizontally. Loganberries can be grown against posts which should be about 6 ft above ground for the canes will not grow longer. Plant in March, 5 ft apart but not too deep and cut back the canes to 3 in. of the base. New canes will then arise and these will bear fruit next season. They are tied in as they grow and this is important for the wood is brittle and easily broken by strong wind. Though loganberries are immune to frost damaging the blossom the plant is not tolerant of cold winds so must be planted in a sheltered place. It also requires full sun, requiring entirely opposite conditions to the blackberry.

Loganberries fruit in August, as do raspberries, the fruit of which they resemble in many respects. Early summer, give the plants a mulch of peat and manure, which will help to retain moisture in the soil. Give another mulch in autumn when the old canes which have borne fruit are cut out near the base and the new canes tied in carefully.

There is now a thornless variety, the canes being free of the sharp hairs of the original form from which it is a 'sport'. If anything, it is more vigorous but in all other respects it is similar.

Propagation is best done by dividing the plants as for raspberries, separating the canes and replanting. Do this in March. Then cut back the canes to 3 in. when new canes will grow to fruit next year. If one or two plants are divided each year, this will maintain the stock in a healthy condition.

Hybrid berries

There are numerous hybrid fruits of value to serve raw, to put in the deep freeze, to make tarts and flans and for preserves. They are all too neglected. They are mostly derived from varieties of blackberry and loganberry (the loganberry is believed to be a red form of the Californian blackberry though from the shape of its fruit, its habit and botanical characteristics it would seem more likely to be a blackberry–raspberry cross, perhaps a natural hybrid).

Boysenberry. An American introduction now rarely seen but it has the vigour of the blackberry and sends out its canes to 10 ft or more. It crops between the loganberry, which it resembles, with its large mulberry-coloured fruits, and the blackberry, ripening its fruit late August and early September.

Laxtonberry. The result of a cross between the loganberry and raspberry, it forms upright canes about 8 ft long and bears long drupes (fruits) like loganberries though 'plugs' like a raspberry. It crops better if planted close to a plantation of the two parent fruits, to assist with pollination, and it ripens in August, at the same time.

Lowberry. An American introduction, it was raised by Judge Logan from a blackberry–loganberry cross and needs the same cultural requirements as the latter, being intolerant of cold winds and hard frosts. It should therefore be grown against a south or west wall where it will send out its thorny canes to at least 10 ft. The fruits are more than 1 in. long, like huge loganberries but shining jet black and with more of a blackberry flavour.

Wineberry. Also known as the Japanese Wineberry as it is of Eastern origin. It was introduced into Europe exactly a century ago though has never been widely planted in Britain. It is well worth growing for its decorative value alone; the canes grow to 8 ft and are clothed in crimson hairs whilst the leaves are silvery white on the underside. It is best grown against a wall or trellis and it fruits on the new wood which is attractive during winter. It fruits at the same time as the loganberry and though not large, the berries are of a deep amber colour, streaked with red and possess a unique wine-like flavour, resembling red grapes. Serve cold, sprinkled with wine or use for preserves. The fruits freeze well.

Youngberry. Another American variety, the canes have moderate vigour, growing 6–7 ft, and it crops on the new wood. The large, round fruits measure up to 1 in. across when grown well and up to 10 lb of fruit may be expected from each plant. The berries mature in July and early August, ripening to a rich purple-black of excellent flavour. There is also a thornless variety, a 'sport', which is more vigorous and crops equally well.

11 Marketing and Presentation

The home grower may find that he has quantities of soft fruits over and above his needs, and where this is the case, he may wish to consider offering his fruits for sale. If there is room, one may grow soft fruits with this in view and build up a profitable concern supplying greengrocers or supermarkets, or selling direct to one's neighbours, who will pay top prices for fruit which has been well grown. But where the home grower usually comes unstuck is in the presentation of the fruit rather than in its culture. Although world-famous paper concerns, such as British Cellophane Limited and the Bowater Paper Corporation, are now devoting considerable expense to bringing to the grower the most modern methods of presentation, the housewife is still confronted with the more perishable fruits, e.g. raspberries and strawberries, served from large baskets into paper bags; the fruit is almost unrecognisable by the time it reaches the home. The housewife today is willing to pay a few pence extra for correctly presented fruit, for she knows that it will not have been crushed either during transportation from the grower or on the way from the shop to her home. Fruit correctly presented will always make extra money, not only because the produce will be in a sound condition but, equally, because in half-pound and one-pound punnets the retailer may make a more attractive presentation. Thus he will prefer well-packed fruit and will pay more for it. The grower will find that competition is always great for quality fruit well marketed. There is now no excuse for failing to market anything but the largest of dessert berries in the most attractive containers. The days when soft fruit was marketed in large baskets, oozing with juice and a mass of stalks and leaves, have gone for ever, and rightly so.

Condition of the fruit

The produce must be clean; fruit splashed with soil is unsaleable. It should be picked only when dry, for wet fruit, however well packed, will quickly deteriorate if travelling long distances. For this reason dessert gooseberries are a more reliable crop to grow than raspberries in the wetter districts of the north Midlands and north-west England. Also, certain varieties of strawberry, i.e. those with a glossy skin and smooth surface, should for the same reason be grown to the exclusion of others. All fruit for sale must be free from blemish and from those untidy stalks and leaves, which the busy housewife hasn't time to remove and which greatly spoil the appearance of well-grown fruit. It is also important to market a fair sample to obtain the confidence of both retailer and housewife, there being nothing more annoying than to obtain well-packed fruit of choice appearance, only to discover small, unripened berries under the top layer. Grade your produce and present it fairly. It is essential to build sound connections, and this should be the aim of every grower who intends to sell his fruit.

The fruit should be packed without crushing but sufficiently firmly to prevent undue shaking during transit. This means packing in a container of just the right size. Small punnets should be sent to the wholesaler in trays of the correct size, perhaps holding a dozen or more, and the punnets should be covered with cellophane paper to keep the fruit clean and prevent pilfering. Half-pound punnets may also be marketed in large chip baskets. Cardboard boxes are not recommended as they lack suitable ventilation. It should be remembered that half-pound punnets of the very highest quality fruit for eating outdoors, or for special occasions, will always make high prices, and are well worth the small additional cost and time entailed.

Great strides have been made recently in the manufacture of special fibre-board punnet trays by

Bowater Fibre Containers Limited, of Stratton Street, London W.1., a subsidiary of the Bowater Paper Corporation. Two sizes are at present being made. One is 18 in. × 14 in. × 3 in. and holds sixteen half-pound punnets or twenty threequarter-pound punnets, a size which is becoming more and more popular for the most choice dessert fruit.

Each tray, made in one piece, is delivered flat and may be assembled in a few seconds by unskilled labour. It is self-locking, with reinforced sides, altogether a great improvement over other existing methods, and the cost will easily be recouped by the resultant superior presentation. Punnets manufactured by the Hartmann Fibre Co., of Finsbury Square, London E.C.2, are equally reliable.

As to the type of punnet to use, the waxed paper punnet travels better than the more common basket punnet. There are numerous types available, each specially designed to give correct ventilation and keep the fruit as cool as possible. It is most important to order packing materials well in advance of the first crop, for nothing is more annoying to the grower than hold-ups in transit when his fruit is abundant and the demand at its greatest.

Market requirements

Before marketing, it is advisable to contact both wholesaler and retailer not only to find the way in which the more famous growers present their fruits, but to discover the exact requirements of those who are to handle your produce. This is all-important, for requirements vary not only in presentation but in varieties preferred in different parts of the country.

It is important to discover the most popular days for marketing your crop, and to try to pick the fruit when most in demand and likely to make top prices. With blackcurrants and gooseberries, this may easily be regulated, but with the more perishable fruits, it is not so easy. Where marketing a considerable distance from your home, it is also necessary to ensure that the fruit will reach the wholesaler with the minimum of delay, for apart from gooseberries, all soft fruit is highly perishable and will begin to deteriorate the moment it is picked. Excessive handling of the fruit should, for this reason, be avoided. The hands should be clean and dry and the fruit picked with care. Any bruising or squashing will cause rapid deterioration. It is advisable to do the packing in a cool room where the fruit can await transportation. A cellar, or a well-ventilated shed built of breeze blocks and in a shady position, will more than repay its cost. The grower should pick the fruit in a slightly under-ripe condition, unless marketing locally. Like flowers, which continue to open after picking, soft fruits continue to ripen. If picked fully ripe, the fruit will have lost its bloom by the time it reaches the shops.

Picking and marketing individual fruits

Blackberry and hybrid berries. Both the loganberry and the earliest varieties of blackberries commence fruiting in mid-July—a useful time, for by then the main flush of strawberries, gooseberries and black-currants is over. Here again, the best dessert fruit should be marketed in quarter-pound, half-pound or one-pound punnets to prevent the berries from crushing. Blackberries will retain their quality on the plants for a considerably longer period than logan-berries, which soon turn a dark crimson and become very soft to the touch. But while the blackberry may be picked when barely ripe, a loganberry not completely ripe will be hard, sour and of little use except for jam-making. The blackberry is also better able to withstand rain; in wet weather the loganberry deteriorates rapidly. 'Extra Selected' loganberries should number not more than 85 to a pound; 'Selected' not more than 128.

Blackcurrants. Blackcurrants are not difficult to pick. There are no thorns, as with gooseberries, no back-breaking stooping as with strawberries, and they make a very much larger berry than either the red or white varieties (though the red American introduction, Red Lake, is rivalling them for size). Though not so necessary as with red and white currants, the blacks will benefit from netting where this can conveniently be done. For home use, the fruit can be picked as soon as it has turned a deep black. Do not pick too soon, or the full sweetness and delicious blackcurrant flavour will be lost.

Never attempt to pick blackcurrants when wet or over-ripe. For this reason they are nothing like so accommodating as the gooseberry. When removing the fruit, try where possible to hold the stalk, rather like picking a bunch of grapes. This prevents the berries being damaged or bruised by the hands. Laxton's Giant and the best fruit should always be marketed in punnets of either quarter-pound or half-pound size and covered with cellophane paper, or in the ready-made 'vision' punnets. Second-quality fruit for jam-making is better marketed in chip baskets or one-pound punnets.

Blueberries. The fruit should be marketed as it is about to turn from crimson to purple, just before it is fully ripe, especially if it has to be transported long distances. Market in half-pound or one-pound waxed punnets.

Gooseberries. For culinary fruit for hotels or for bottling, the fruit may be gathered before it is fully mature and marketed in six-pound baskets as there is no fear of quick deterioration. Dessert fruit should be allowed to become fully ripe and then marketed in half-pound or one-pound punnets to be consumed in the open. Never mix colours in the same punnet, though punnets of several varieties and colours may be marketed in the same consignment.

The skins should be firm and free from blemishes, and the punnets should be marketed as 'Selected Dessert', containing not more than 150 berries to the pound, though for those monster varieties this may be as few as 30–40 to the pound. With 'Selected' Leveller, a pound should not exceed 28 berries.

Raspberries. Raspberries are not easy to pick and market in perfect condition. A period of good weather is desirable when cropping and ample time should be allowed for regularly looking over the canes. Over-ripe fruit will be useless for anything and almost impossible to pick. Some varieties come away from the core better than others, which is of course a great advantage. Lloyd George, Malling Notable and Malling Admiral are excellent in this respect, and may be picked without any bruising or squashing. Owing to the fact that raspberries, more than any other soft fruit, tend to become squashed so easily, it is never wise to retail in too large amounts: half-pound punnets are better than one-pound size for this reason. Even so, fruits at the bottom may become squashed if picked during a rainy period and kept for any length of time in the punnets. But the same marketing difficulties beset the foreign growers, who therefore send few raspberries to English markets, thus the price of the fresh fruit is always fairly constant.

For 'Extra Selected' fruit, there should not be more than 128 berries to a pound, and they should be free from dirt, mildew and grubs. 'Selected' fruit should contain not more than 256 berries to a pound.

Red Currants. It has already been said that the great worry with red currants is fear of attack from birds, so those bushes which cannot be covered should have the fruit removed as soon as it is ripe, or even a little earlier. Pick the fruit on the strig (as the stem is called), for red currants do not improve with handling, and market in one-pound or half-pound punnets. Those contemplating growing red currants will be well advised to choose only those varieties that yield a large fruit; small seedy berries are not wanted by anyone. As with blackcurrants, strigs containing few fruits give the pack an untidy appearance and are not wanted.

Rhubarb. Colour and length of stem will determine when to pull. To obtain depth of colour in forced rhubarb, admit light for twenty-four hours whenever possible. Care should always be exercised in pulling the stems, whether the rhubarb is being forced or is growing in the open, for they break so easily and broken stems are unsaleable. Hold them firmly at the base and pull in an upward direction. Indoor grown rhubarb should be marketed as stems of even length, one thick stem with one of medium size, or two thin stems, being fastened together with a rubber band at the bottom and at the top below the leaves. Broken and unduly bent or twisted stems should be used in the home.

Outdoor grown rhubarb should be made into larger bunches, containing about six sticks tied top and bottom with raffia. The large leaves should be cut neatly off leaving only a very few at the top.

Strawberries. Like gooseberries, strawberries are a satisfying crop to pick, for they weigh heavy and professional pickers can pick up to twenty pounds in an hour. The fruit should be dry and quite clear of soil splashing. It should be marketed when a deep pink colour, rather than crimson, yet should bear no tinge of green. Strawberry leaves placed round the punnet will improve the appearance. The fruit should be quite even in size. 'Extra Selected' fruit should weigh not less than three-quarters of an ounce each fruit; 'Selected' not less than one-sixth of an ounce each. A bright, clean, orange-scarlet fruit will always catch the buyer's eye better than fruit of a dull crimson colour. Any soil clinging to the fruit should be carefully removed with a camel-hair brush. Half-pound punnets are the most popular size.

Marketing for deep freezing and canning

Because the processors require the fruit in bulk, it must not be thought that it can be marketed in rough condition. The fruit should be picked just before it is fully ripe, and sent in six-pound chip baskets covered with cellophane or cardboard. It is important for it to reach the processor in the quickest possible time, and as regards cleanliness and quality, the standards are as high as for the fresh-fruit market. Only good-quality fruit is required for freezing and canning. So important is it to freeze the fruit as soon as possible after picking, that many of the freezers are now growing their own requirements of certain fruits.

Upon arrival at the factory, the fruit is subjected to a temperature of 60° below freezing, in either sugar or syrup. The fruit is not kept waiting for more than an hour at the most, and for this reason, factories are located near the major growing districts. Both freezers and canners pay more for fruit which has been plugged, i.e. free of the stem and plug which holds the fruit to the stem. The difficulty of removing the plug from certain varieties makes them of little use for canning. As many as one thousand tons of strawberries are used annually for freezing, whilst raspberries and loganberries are rapidly gaining in popularity, so it will repay any grower to consider this market with care.

Index